# Exploring BIOLOGY in the Laboratory

## CORE CONCEPTS

Murray P. Pendarvis
*Southeastern Louisiana University*

John L. Crawley

**Customized for Virginia Western Community College**

**Biology 101**

925 W. Kenyon Ave., Unit 12
Englewood, CO 80110

www.morton-pub.com

Printed in the United States of America

10 9 8 7 6 5 4 3 2 1

ISBN-10: 1-61731-825-6
ISBN-13: 978-1-61731-825-2

# Biology Lab Rules & Regulations

*Below is a list of general lab rules and regulations that need to be followed at all times.*

## 1. General

a. Be on time; if you are late, you may be prohibited from participating in the exercise and be marked absent.
b. Conduct yourself responsibly and be alert at all times.
c. No horseplay in the lab.
d. Read the laboratory exercise before coming to the lab, which will better prepare you for a safe, enjoyable and informative session.
e. During the lab, read the whole procedure before you start any experiment.
f. Follow instructor's directions and, if not sure, ask questions.
g. Eating, drinking, smoking, applying cosmetics or lip balm, and handling contact lenses are strictly prohibited in the laboratory.
h. Use caution when handling glassware and when heating liquids.

## 2. Dress Code

a. No open-toed shoes allowed in the lab.
b. Long hair must be tied at all times.
c. Goggles and gloves must be used for certain exercises and instructed by the procedure.

## 3. First Aid

a. Report all accidents to instructor.
b. Assume that all chemicals are corrosive and toxic. If your skin gets in touch with any chemical, rinse thoroughly and inform the instructor immediately.
c. Know the location of the fire extinguisher, fire blanket, eye wash, first aid kit and broken glass box.

i. Fire extinguisher is located: ____________________
ii. Fire blanket is located: ____________________
iii. Eye wash is located: ____________________
iv. First aid kit is located: ____________________
v. Broken glass box is located: ____________________

## 4. End of lab session

a. Wash all glassware and material that you used.
b. Return all material used back to designated spot.
c. Microscopes must be cleaned and stored properly as directed by instructor.
d. Wipe the bench with disinfectant.
e. Wash your hands before you leave.

## Discussion Questions

1. Why can't you drink or eat in the lab?

2. Why can't you participate in the lab if you come in late?

3. How can reading the lab prior to coming be helpful?

4. Why read the whole procedure, all steps, before starting an experiment?

5. Why should you clean the bench before leaving?

6. Why let the instructor know if you break something?

7. Look around the room, what safety equipment do you find?

8. Look around the lab and find each of the following:

    1. Glassware __________
    2. Drying rack __________
    3. Dissection equipment __________
    4. Markers __________
    5. Tape __________
    6. Paper towel __________
    7. Soap __________
    8. Disinfectant __________
    9. Alcohol __________
    10. Microscope supplies __________
    11. Balances __________
    12. Trash __________
    13. Broken glass box __________

# Contents

# The Scientific Method

1

***Before coming to lab***

- ☐ Read the lab
- ☐ Answer the pre lab questions

## PRE LAB QUESTIONS

1. List the steps of the scientific method?

2. What is the difference between dependent and independent variable?

3. What type of organism will we use in today's lab?

4. What is a hypothesis?

5. What are the two most common graphs used to summarize data?

6. What are three things that are important to include in a graph?

7. If you could design an experiment to answer a question, what would your question be?

## Purpose of the lab

The purpose of these laboratory exercises is to introduce the student to the scientific method and the design of experiments. In addition, students will be introduced to data graphing and reporting. Students will design, conduct and report results of an experiment.

### Introduction

Biology is the study of living organisms and is a form of science that is usually based on observation. It is very important to base science on objective evidence and not ideas or opinions. Therefore, scientists usually use a very organized process called the **scientific method** to find explanations for the world around them. The scientific method is a logical and organized process that includes a series of steps that are essential to ensure that conclusions are objective and verifiable. Following is a flowchart of the different steps of the scientific method.

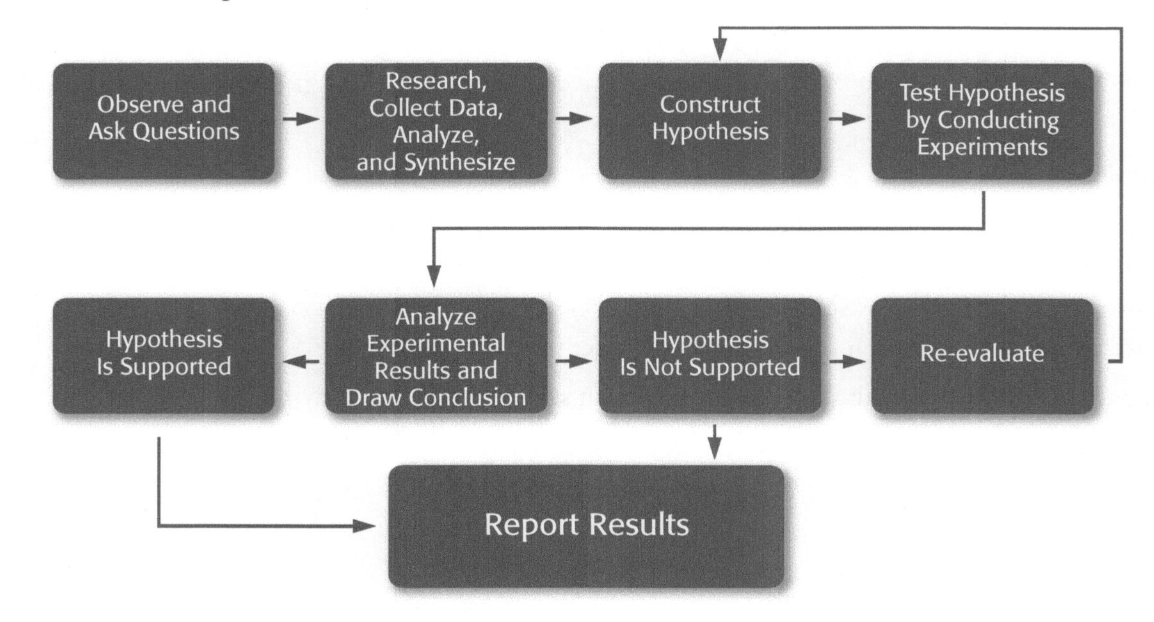


Below is a description of each of the steps of the scientific method.

1. The first step of a scientific method is **observations.** Usually a biologist that is interested in a particular topic will be very curious and spends time reading about it and observing it. This step involves researching the topic by reviewing what is already known about it.
2. Observations usually lead to questions. The second step of the scientific method is to **ask a question.** This is what you are attempting to answer.
3. The third step is constructing a **hypothesis.** It is a possible answer to the question stated that is based on the research and previous knowledge of the topic under investigation. Also called an educated guess.
4. A **prediction** follows that is based on the hypothesis. The prediction usually takes the form of "if...... then......" A prediction states the expected results if the hypothesis is correct.
5. **Test** the hypothesis by either conducting a controlled experiment or making numerous observations when experiments cannot be made.
6. The final step in the scientific method is the **conclusion**, which summarizes the results of the experiment and relates them to the hypothesis.
   a. The results support the hypothesis and therefore you accept your hypothesis.
   b. The results do not support the hypothesis and therefore you reject the hypothesis and form a new hypothesis to test.

You use the scientific method every day without realizing it. Below are three everyday observations: work with a partner to address one example using the scientific method.

- Your dog is not eating.
- Your car would not start.
- A plant in a pot is dying.

Apply the scientific method to one of the above examples.

1. Observation:
2. Question:
3. Hypothesis:
4. Prediction:
5. Experiment:
6. Conclusion:

Scientists usually use the scientific method to answer more complex questions such as:

- Will reducing $CO_2$ emissions reduce global warming?
- Can a drug treat a certain disease?
- Will vaccinating against human papillomavirus protect from cervical cancer?

Work with a partner to come up with three other things that scientists may be studying.

1.
2.
3.

## Designing Experiments

Designing the appropriate experiment is the most important part of the scientific method. Experiments must be objective, verifiable, repeatable, and designed to test one hypothesis at a time. A good experiment must:

- Be controlled.
- Be fair and unbiased.
- Be randomized.
- Be easy to repeat.
- Test the hypothesis.
- Usually test one variable at a time.
- Have repeated trials to reduce errors.

Experiments have several variables where a **variable** is any factor that can change or vary and can have an impact on the experiment. There are three kinds of variables:

**1. Independent variable**

- The variable that is controlled and manipulated by the scientist.
- The variable that is being tested.
- Experiment should have only one independent variable.

2. **Dependent variable**
   - Affected by the independent variable.
   - Variable that is being measured.
   - The data that is being measured.
3. **Control**
   - Factor that is not changed and kept constant during the experiment.
   - Several factors that are held constant.

In most experiments two groups are studied: the **control group**, which is the group that is not exposed to the independent variable, and the **experimental group**, which is exposed to the independent variable.
Let's use some examples to practice these concepts. **Choose one of the examples** listed below and answer the following questions.

## Example 1

You were told that exercising helps you lose weight. You don't believe these claims and want to test for yourself.

1. Question:
2. Hypothesis:
3. Prediction:
4. How would you conduct the experiment?
   a. Independent variable
   b. Dependent variable
   c. Control variables
5. What would you conclude?

## Example 2

You are having problems with your plants; they are not growing as fast as you would like. A neighbor suggests that adding dish detergent will make them grow faster. You are skeptical and as a scientist want to test that for yourself.

1. Question:
2. Hypothesis:
3. Prediction:
4. How would you conduct the experiment?
   a. Independent variable
   b. Dependent variable
   c. Control variables
5. What would you conclude?

### Example 3

Suggest an observation or question and answer the following questions.

1. Question:
2. Hypothesis:
3. Prediction:
4. How would you conduct the experiment?
    a. Independent variable
    b. Dependent variable
    c. Control variables
5. What would you conclude?

## Reporting data: graphing

There are several ways that scientists report results of experiments: for example, tables and graphs. Graphs are very helpful since they present data in a pictorial format that is easy to understand. A graph should make sense without any additional explanation. For that reason there are parts of the graph that are absolutely necessary. These are:

1. Title.
2. Legend.
3. X-axis label with units, if applicable.
4. Y-axis label with measurements, if applicable.

## Types of graphs

There are several types of graphs. The type of graph one uses depends on the type of data and what makes sense in proving the point and expressing the idea that is being conveyed. Three of the most common types of graphs are:

1. **Bar graphs** are used to compare values from two or more categories, when the independent variable is discrete (no points between data). Examples:

   Average height for males versus females.

   Effect of fertilizer on plant growth.

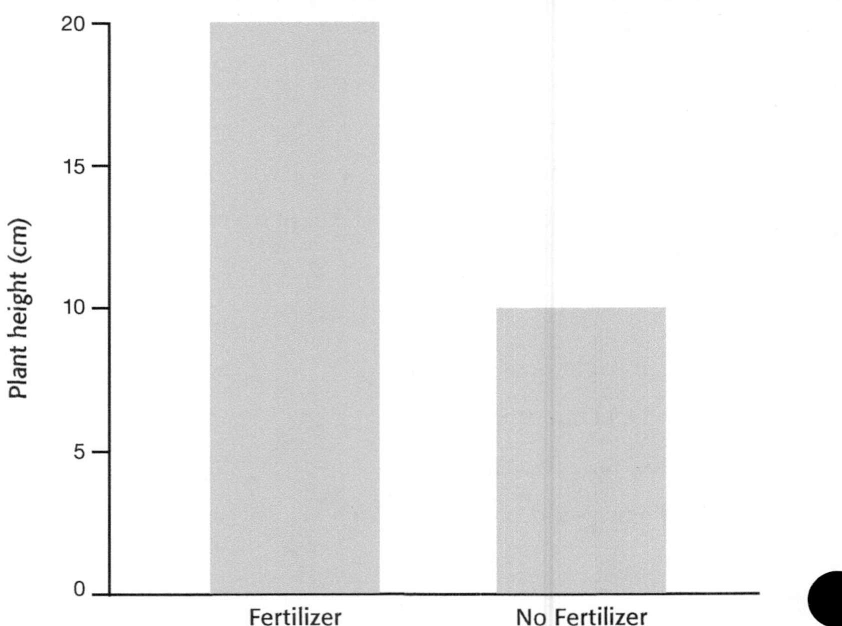

2. **Line graphs** are often used to show data that is part of a continuous process. Examples:

   Weight of a child over a three year period.

   How temperature affects plant growth.

   How temperature changes during the day.

   Average weight of boys over first 5 years.

3. **Pie graphs** are used to show the contribution of different categories to a whole. Each category is represented by a wedge, and the size of the wedge is proportional to the percent of that category. Examples:

   Percentages of the race/ethnicity of public school students across the U.S.

   Percentages of different races of AIDS patients.

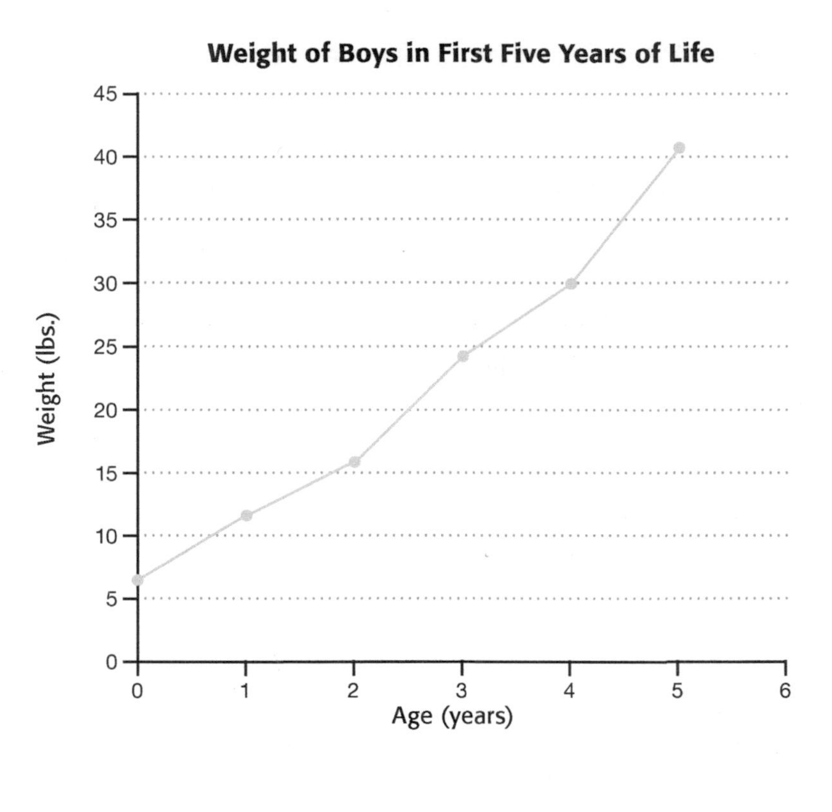


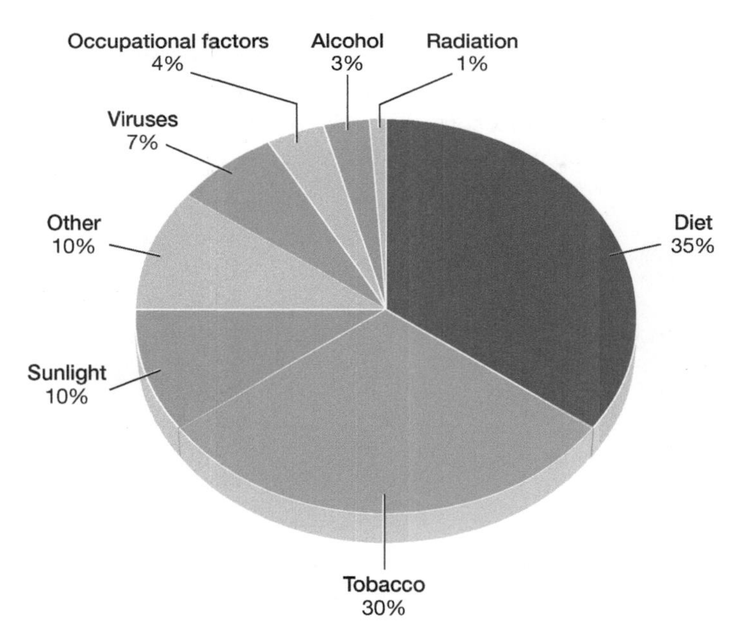


## Helpful points in constructing graphs

1. The independent variable goes on the x-axis.
2. The dependent variable goes on the y-axis.
3. If you are drawing a graph, use a pencil, a straightedge and graph paper.
4. If you are using a computer for constructing a graph, make sure the graph is correct and makes sense.
5. Label each axis properly.
6. If an axis includes the results of a measurement, make sure to include the units used to measure (e.g., degrees C, millimeters, hours, kilograms, liters).
7. Don't put a number by every tick mark (unit of measurement) on the axes; that just clutters an axis.
8. Units on the scale should be at regular intervals.
9. Scales often start at "0," but do not have to. Choose a scale that does not waste space and makes sense. Choice of scale can have a drastic effect on the appearance and efficiency of a graph.
10. Give the graph a title that specifically describes what information is in the graph. **Do not** use vague titles like these: "Biology Lab Exercise 1," "Cell Lab."
11. Graphs must stand alone.

# Scientific Method Experiment

## The Ubiquity of Bacteria

The objective of this part of the lab is for students to practice the scientific method by designing and conducting an experiment, and then summarizing their findings. Two experiments are listed below; depending on time, you might be conducting only one or both.

Bacteria are ubiquitous, unicellular, prokaryotic organisms that make up most of the life on earth. Let's review these terms:

A. What does ubiquitous mean? ______________

B. What does unicellular mean? ______________

C. What does prokaryotic mean? ______________

Surprising to most students is the fact that the vast majority of bacteria are NOT harmful, but in fact many are actually helpful. Bacteria can be beneficial to us personally, and they also contribute significantly to the environment around us. Please discuss and list several ways that bacteria can be helpful to us.

1. ______________
2. ______________
3. ______________
4. ______________
5. ______________

**Fun Fact**

There are 10 times MORE bacteria living in you and on you than there are human cells! Further, estimates show that as much as 2–9 POUNDS of our mass are made up of bacterial cells! The belly button has been found to be home to 1,400 different species!!!

Bacteria can also be *pathogenic*, or able to cause disease. Bacteria can cause disease in plant, animals and, of course, in humans. Historically, many deaths have been caused by bacterial diseases such as tuberculosis, cholera and, of course, bubonic plague. List some bacterial diseases that are common in our population today:

1. ______________
2. ______________
3. ______________
4. ______________
5. ______________

WHILE EACH OF THESE DISEASES IS IMPORTANT AND CAN HAVE A MARKED IMPACT ON HUMAN HEALTH, IT IS IMPERATIVE TO REMEMBER THAT MOST BACTERIA ARE NOT PATHOGENIC—ONLY THE MINORITY ARE CAPABLE OF CAUSING DISEASE.

## Culturing Bacteria and Exploring Bacterial Ubiquity

Much that we know about bacteria is because we have learned to grow bacteria in the laboratory. Robert Koch, a famous German microbiologist who lived in the 1800s, developed and perfected many of the techniques for growing bacteria that we still use today. In fact, he was so enthusiastic about learning to culture microbes that he set up a lab in his own home, and for holidays, his wife always gave him laboratory equipment as presents!

Just like all living organisms, bacteria require a nutrient source so we have to provide bacteria with nutrients when growing them. Bacteria have many different foods that they like, and not all bacteria like the same food (i.e., some are very "picky eaters!"). Today we are going to use a food source for the bacteria that most of them do like. Tryptic Soy Agar (TSA) is the growth medium that you will be using for this ubiquity experiment. This is a general purpose medium that will grow many different types of bacteria and fungi. The plates that you will receive have been prepared using aseptic technique, which means that the plates are sterile until you open them—nothing will grow unless you put microbes on them!

The first thing to do when you receive your TSA plates is to label them correctly. This means that you label the plates on the same side where the agar is poured (not on the lid), and you want to label them using small print so that you do not obscure the organisms you are trying to grow—you will want to look at them next week. You will label the plates with your initials, the date and the sample—your instructor will show you a correctly labeled plate.

When you are ready to collect a sample, you will use a sterile cotton swab to swipe the area you are testing for bacterial growth. If you are collecting a sample from a dry area, such as a desktop, you should moisten the swab with sterile water first. Then swipe the area back and forth 3-4 times, then quickly open the sterile TSA plate, and while still holding the lid, move the swab back and forth across the agar with a gentle touch—do not push the swab into the agar—just swipe it across the top. The plate can be divided into sections to put more than one sample on the same plate as shown below:

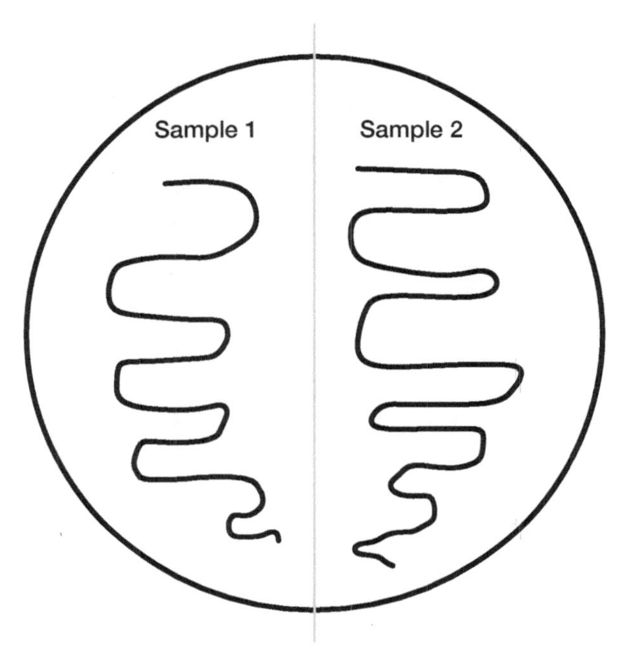


Always use a clean swab for each sample, and place the swabs in the biohazard bags when you have finished. Replace the lids on the plates as soon as possible, and don't lay the lids on the table—you want only the sample you placed on the agar to grow—you do not want to introduce unwanted organisms. You can use the lid as a sort of shield to cover the plate while it is open.

After you have placed your sample onto the properly labeled plate, you will invert the plate and place the plate in the basket at the front of the room. The plates are inverted before incubation so that condensation does not form on the surface of the agar and cause the different bacteria to mix. These plates will be incubated for 24 hours at 37° Celsius and then at room temperature until observation.

When observing your plates, you will see "dots" of bacteria—each of these dots is called a colony. Each colony contains millions of bacterial cells, all of which are usually genetically identical because a colony derives from one bacterial cell. The colonies may look different from each other—they may be different sizes, color, shapes and textures, for example. Colonies that look different from each other likely represent different bacterial species. You can estimate how many different species are present from a given sample just by examining this colony morphology. You will be working with your group of lab partners to answer some basic questions about where we may find bacteria in the environment around us. These questions are:

1. Do you think there are more bacteria in your mouth or in your ear?
2. Do you think there are more bacteria on the toilet seats from the men's room or the women's room?
3. Do you think there are more bacteria in the soil or on the bottom of your shoe?
4. Are there more bacteria on the desk or on a cell phone?

Each group should obtain 1 TSA plate. Working as a group, please use the scientific method to develop an experiment to address one of the following questions. Your instructor will assign one of the scenarios to your group. Fill out the following questions that correspond to your experiment. Observations will be made and the data recorded during the next lab period.

1. Mouth or Ear?
   a. Observations:
   b. What is your hypothesis?
   c. How will you design your experiment using only one plate?
   d. Results: Please draw your plate.

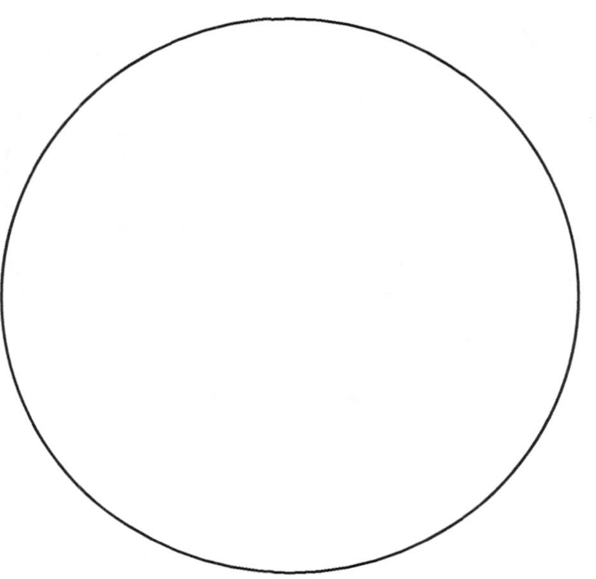

   Approximately how many different species do you see on this plate?

   e. Conclusion—Was your hypothesis supported or refuted? Explain.

2. Men's Room or Women's Room?

   a. Observations:

   b. What is your hypothesis?

   c. How will you design your experiment using only one plate?

   d. Results: Please draw your plate.

   Approximately how many different species do you see on this plate?

   e. Conclusion—Was your hypothesis supported or refuted? Explain.

3. Soil or Shoe?
   a. Observations:

   b. What is your hypothesis?
   c. How will you design your experiment using only one plate?

   d. Results: Please draw your plate.

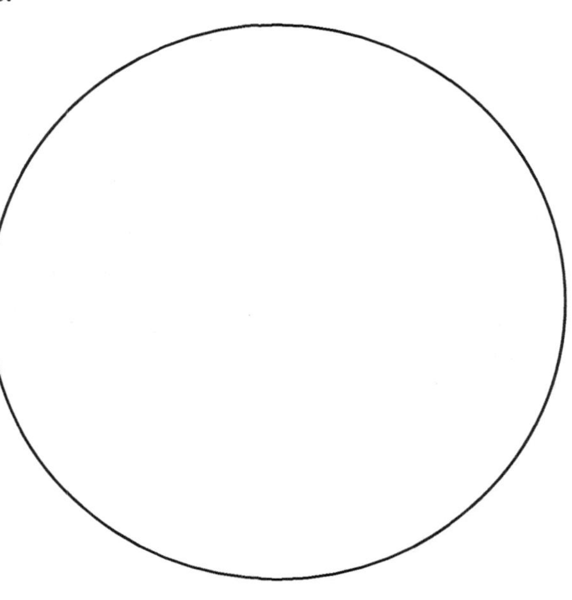

   Approximately how many different species do you see on this plate?
   e. Conclusion—Was your hypothesis supported or refuted? Explain.

4. Desk or Cell Phone?
   a. Observations:

   b. What is your hypothesis?
   c. How will you design your experiment using only one plate?

   d. Results: Please draw your plate.

   Approximately how many different species do you see on this plate?
   e. Conclusion—Was your hypothesis supported or refuted? Explain.

# Scientific Measurements

## Purpose of the lab

The purpose of these laboratory exercises is to introduce the students to basic scientific measurements and familiarize them with common laboratory equipment used to measure length, mass, volume and temperature using the metric system.

## Introduction

Precise, accurate measurements are often needed when observing and investigating the living world. The American system is still in use in the USA for everyday measurements. However, in the scientific world and in everyday measurements almost all over the world, the universal system that is in use is the metric system, also called International System of Units. This system is easy, precise and convenient once learned. It is based on the number 10, which makes conversions within the system very easy.

Why the metric system?

1. Nearly the entire world, except the United States, uses the metric system.
2. Metric system is used exclusively in science—therefore, to better understand science we need a better understanding of this system.
3. The metric system is *much* easier than the American system. All metric units are related by factors of 10, which makes it easier to convert between units.

What makes the metric simple and easy to use is the fact that there is only one unit of measurement (base unit) for each quantity or type of measurement (length, weight....). The following table shows the metric base units as well as the American units for different measurements.

| Measurement | American | Metric base units (symbol) |
|---|---|---|
| Length | Inch, feet, mile | Meter (m) |
| Weight | Pounds, ounces | Gram (g) |
| Volume | Quart, gallon | Liter (L) |
| Temperature | Degrees Fahrenheit (°F) | Degrees Celsius (°C) |

Within the metric system, prefixes are used to designate units smaller or larger than the base unit. The same prefixes are used for all measurement types. The following table shows some commonly used metric prefixes.

| Prefix | Symbol | Equivalent | Factor |
|---|---|---|---|
| Nano | N | 0.000000001 | $10^{-9}$ |
| Micro | µ | 0.000001 | $10^{-6}$ |
| Milli | M | 0.001 | $10^{-3}$ |
| Centi | C | 0.01 | $10^{-2}$ |
| Deci | D | 0.1 | $10^{-1}$ |
| base unit | | 1 | $10^{0}$ |
| Deka | Da | 10 | $10^{1}$ |
| Hecta | H | 100 | $10^{2}$ |
| Kilo | K | 1,000 | $10^{3}$ |
| Mega | M | 1,000,000 | $10^{6}$ |

Prefixes are added to the base unit to designate the different metric units. Following is a table with different measurements and the metric units used.

| Measurement | Unit | Symbol |
|---|---|---|
| Length | Millimeter | mm |
| | Centimeter | cm |
| | Meter | m |
| | Kilometer | km |
| Mass | Milligram | mg |
| | Gram | g |
| | Kilogram | kg |
| Volume | Milliliter | mL |
| | Cubic centimeter | $cm^3$ |
| | Liter | L |
| Temperature | Degrees Celsius | °C |

## Metric measurement—Length

**Length** is the distance between two points that can be measured by a ruler. The basic metric unit is the **meter.** Below is a comparison between the metric system and the American system.

| American system | Metric system |
|---|---|
| 12 in. = 1 ft. | 10 mm = 1 cm |
| 3 ft. = 1 yd. | 100 cm = 1 m |
| 1760 yd. = 1 mile | 1000m = 1 km |

English-Metric conversions:

- 1 inch = 2.54 cm
- Length is usually measured by a ruler or meter stick.

## Activity 1

Obtain a meter stick and answer the following questions.

1. One meter equals how many inches? ____________
2. How many cm in a meter? ____________
3. How many millimeters in a cm? ____________
4. What is the height of the lab bench in cm? ____________
5. Convert 250 cm to km. ____________

## Metric measurement—Mass

**Mass** is the quantity of matter in a certain object. The basic metric unit is the **gram.** Many times the terms mass and weight are used interchangeably; however, they are different because weight is dependent on the force of gravity acting on the object. For example, the mass of a certain object is the same on earth and on the moon; however, the weight is much less on the moon because the moon's force of gravity is less than that on earth. Below is a comparison between the metric system and the American system.

| American system | Metric system |
|---|---|
| 437.5 grains = 1 oz. | 1000 mg = 1 g |
| 16 oz. = 1 lb. | 1000 gm = 1 kg |
| 2000 lb. = 1 short ton | 1000 kg = 1 metric ton |

### English-Metric conversions:

- 453 g = 1 lb.
- 2.2 lb. = 1 kg

To measure mass, either a triple beam balance, which has three horizontal beams to measure an object's mass, or a digital balance, is used.

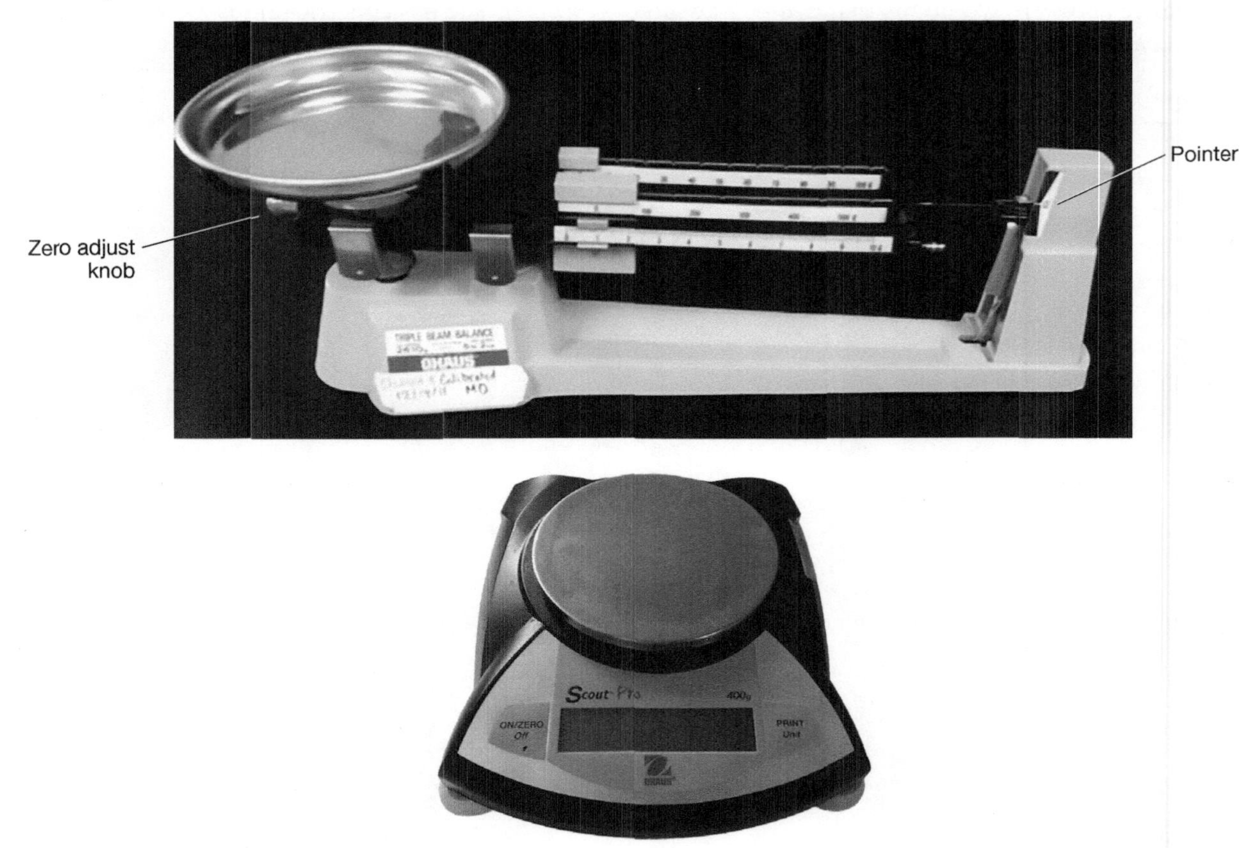


***Using the triple beam balance***

1. Empty the pan on the balance and make sure it is clean.
2. Move the three sliding masses all the way to the left.
3. Check on the right to ensure that the **pointer** aligns with the mark on the right-hand side; this means the balance is calibrated. If it does not, calibrate it by moving the **zero adjust knob** under the pan until they are aligned.
4. Place the object to be measured on the pan, which will cause the pointer to move above the zero line.

5. Move the 100-gram sliding mass on the beam one notch at a time until the pointer on the right end drops below the zero mark. Then move the sliding mass back one notch.
6. Move the 10-gram sliding mass on the beam, one notch at a time, until the pointer on the right end drops below the zero mark. Then move the sliding mass back one notch.
7. Repeat with the 1-gram sliding mass.
8. The mass of the object can be determined as the sum of the masses indicated by all three beams.

## Activity 2

Obtain triple beam balance and answer the following questions.

1. What is the mass of your lab manual? ______
2. What is the mass of a starfish? ______
3. Convert 250 g to mg. ______
4. Convert 0.45 g to kg. ______

## Activity 3

Use a digital balance to weigh 15 grams of beans into a weigh pan and answer the following question.

1. How did you weigh the beans? ______

2. What is the weight of three plastic rulers? ______

### Metric measurement—volume

**Volume** is the space occupied by an object or fluid. The basic metric unit is the **liter**. Below is a comparison between the metric system and the American system.

| American system | Metric system |
|---|---|
| 57.75 $in^3$ = 1 qt. | 1 $cm^3$ = 1 mL |
| 4 qt. = 1 gal. | 1000 mL = 1 liter |

English-Metric conversions:

- 3.79 liters = 1 gal.

Several devices are used to measure volume, such as the beaker, Erlenmeyer flask, graduated cylinder and pipettes shown below.

*Reading a volume*

Because water molecules are cohesive, they stick to each other and to the sides of the container, forming a **meniscus.** When reading the volume of a fluid, you must use the lowest margin of the fluid level of the meniscus as shown.

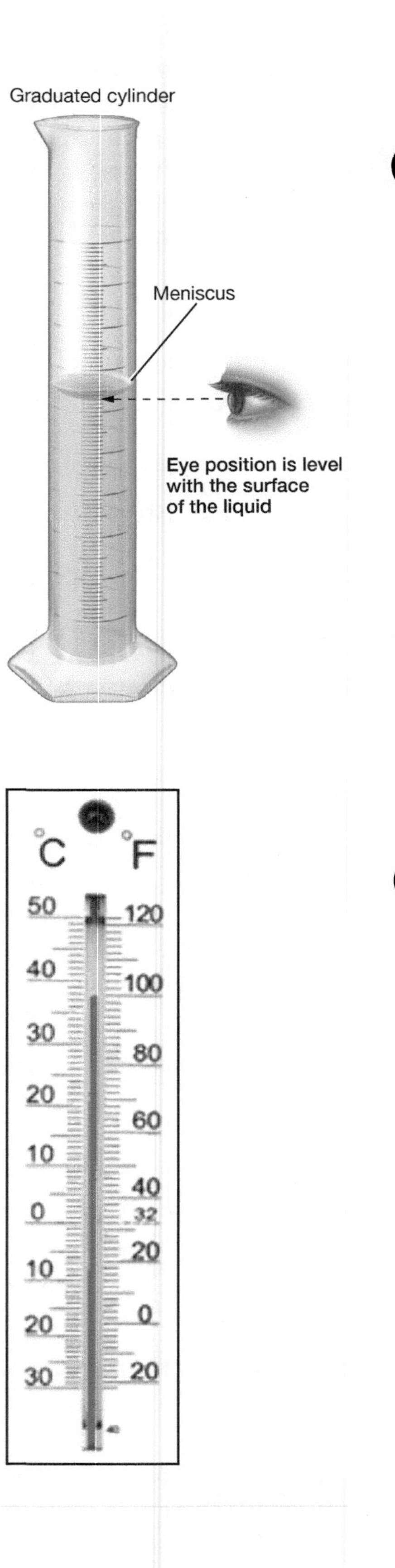


## Activity 4

Obtain a beaker, a graduated cylinder, an Erlenmeyer flask and a pipette, and answer the following questions.

1. What is the volume of water in the distilled water bottle on your desk?

2. Convert 200 mL to L.
3. Convert 350 L to mL.
4. Which of the lab glassware would be most accurate to measure 36 mL?

## Metric measurement—Temperature

Temperature measures the average kinetic energy of molecules. In other words, it is a measure of how hot or cold an object is. Below is a comparison between the metric system and the American system.

| American system | Metric system (°C) | Water |
|---|---|---|
| 32° | 0° | water freezes |
| 68° | 20° | room temperature |
| 98.6° | 37° | body temperature |
| 212° | 100° | water boils |

English-Metric conversions:

To convert between American and metric systems the following formulas are used.

$$T(°C) = [T(°F) - 32] \times 5/9$$

$$T(°F) = T(°C) \times 9/5 + 32$$

To measure temperature a thermometer is used.

## Activity 5

1. Convert 48 °C to °F.
2. Convert 48 °F to °C.

# Metric System Worksheet

Work in pairs to answer the following questions.

## A. Length

a. Use the meter stick to **measure** the length, width, and height of your laboratory table or desk **in cm** and convert to other units to fill the table.

| Lab table measurements | | | |
|---|---|---|---|
| **Measurement** | **m** | **cm** | **mm** |
| Length | | | |
| Width | | | |
| Height | | | |

b. Use a ruler to **measure** the length and width of a test tube **in cm** and convert to other units to fill the table.

| Test tube measurements | | | |
|---|---|---|---|
| **Measurement** | **m** | **cm** | **μm** |
| Length | | | |
| Width | | | |

## B. Weight

a. Use a three beam balance to determine the mass in g of each of the following and convert to fill the table.

| Mass measurements | | | |
|---|---|---|---|
| **Measurement** | **mg** | **g** | **kg** |
| 50 mL beaker | | | |

## C. Volume

a. Use the appropriate lab device to measure the volumes and answer the questions.

i. Fill a test tube to the rim with water, empty into an appropriate cylinder and record the volume in mL.

ii. Fill a 50 mL beaker to the rim, empty into a graduated cylinder and record the volume.

iii. Use a disposable pipette to transfer 10 mL, 1 mL at a time, into a graduated cylinder. **How accurately did you pipette the liquid?**

Record the volume in mL.

## D. Temperature

a. If time permits, take a thermometer and go outdoors for 5 minutes. Determine the temperature in °C and convert to °F. Or check the local weather on the Internet. Show your work.

b. Convert 50°F to °C. Show your work.

# Microscopy and Cells

2

***Before coming to lab***

- ☐ Read the lab
- ☐ Answer the pre lab questions

## PRE LAB QUESTIONS

1. What is a microscope and why do we need to use them in biology?

2. What are the two main kinds of light microscopes?

3. What is depth of field?

4. What are the two main types of cell?

5. What is the plant cell you will use today?

6. What is the animal cell you will use today?

7. Who coined the term cell?

8. Name one photosynthetic bacterium.

9. Name one heterotrophic bacterium.

10. What are protists?

## Purpose of the lab

The purpose of these laboratory exercises is to introduce the student to the basic concepts of light microscopy and to allow the student to practice using two types of light microscopes: the dissecting microscope and the compound light microscope. Further, the student will become familiar with organisms representing the domain Bacteria and the domain Eukarya.

# Part 1: Microscopy

## Introduction

You will be using microscopes throughout this semester, and in many other science classes you take during your education. Learning to use the microscope correctly makes the labs much more interesting and reduces frustration. By using these microscopes, you will be introduced to a world of beautiful, intriguing organisms that you could not otherwise see!

**Background information:** The field of microscopy began in the 1600s, namely with two hobbyists, **Robert Hooke** and **Antoni von Leuweenhoek**. Robert Hooke, most famously, viewed thin slices of tissue from a cork tree under a handmade microscope. When he viewed this plant tissue, he noticed that it was divided into smaller sections, which reminded him of the rooms where monks lived in the monasteries in his native Austria. These rooms for the monks were called "cells," and Robert Hooke used this term to describe the small sections of plant tissue. He was the first person to use the term "cell" in a biological context and his observations laid the foundation for the development of the **cell theory**. You will now spend almost an entire semester studying the cell! Hooke went on to view many specimens with his microscopes and published drawings of his specimens in a book entitled *Micrographia.*

Antoni von Leuweenhoek was a tailor by trade, and was very interested in the fabrics from which he made his clothes. In order to be able to better discern the quality of the fabric he was using, he began to grind glass lenses in order to see the cloth. His lenses were excellent and provided at least an 800-fold magnification. He became very interested in using these lenses, and quickly began looking at things other than cloth. He viewed pond water, minerals, fossils, and even plaque from his own teeth with his microscopes. In all of his samples, he observed living, often moving organisms, which he hired an illustrator to draw for him. His now famous drawings indicate that he was the first person to observe microorganisms such as bacteria and protists. Bacteria are single-celled prokaryotes, and protists are eukaryotes that do not belong to the fungi, plant or animal kingdoms. He called the living organisms "animalcules" and sent many letters to the Royal Society of London announcing to the public that there were organisms alive that were too small to be seen with the naked eye! Try to imagine how this changed the way the people of this time viewed their world—they had never imagined this before and had no idea that organisms that they could not see existed!

In this lab, you will be using various prokaryotic and eukaryotic cells and organisms to learn basic microscopy techniques. Please familiarize yourself with these organisms as you will be asked to identify them later. ***Further, as you learn to use these microscopes today, please understand that they are very expensive pieces of equipment and must be treated accordingly. Please follow all procedures when using the microscope, and clean and store the microscope correctly when the lab is finished.***

**PLEASE NOTE THAT ONLY LENS PAPER IS TO BE USED TO CLEAN THE MICROSCOPES.**

There are two types of microscopes that you will be using this semester: the dissecting scope and the compound light microscope.

### The Dissecting Microscope

The dissecting microscope, also called a stereoscope, is used for viewing specimens that are too large to fit onto a microscope slide. While the magnification is not as great as with the compound microscope, these scopes are easy to use and provide great details on "large" specimens. Light is very important when using any microscope, so please note that there are two light controls on the microscopes, one for a top light and one for a bottom light. Also note the focus knob and the magnification knob on the top.

## Activity 1

Activity 1 is designed to teach students to use the dissecting microscope. When using the dissecting scope, you may adjust magnification, focus and the light source. Your instructor will go over each of these with you.

1. Observe and draw Rotifers.

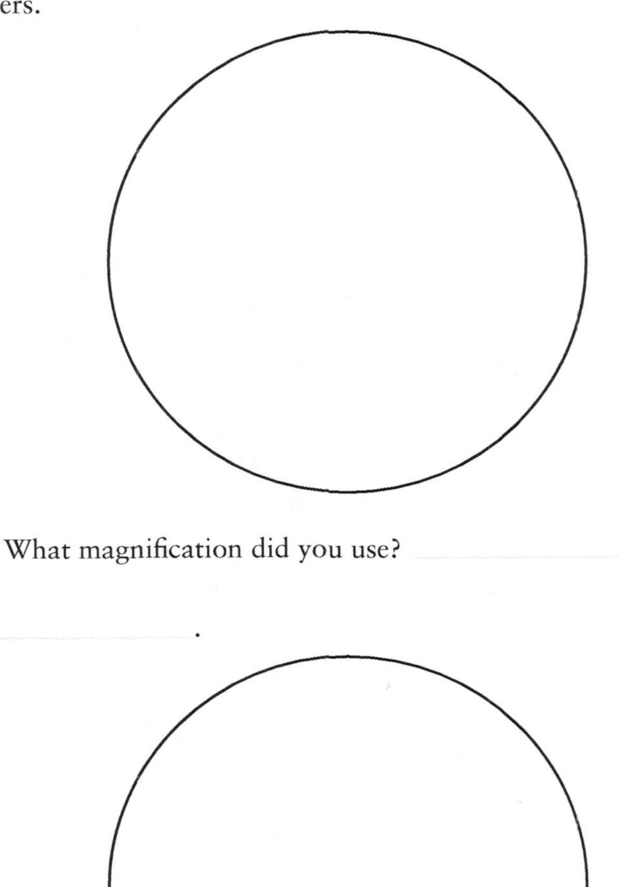

What magnification did you use? ______________

2. Observe and draw ______________.

What magnification did you use? ______________

### Compound Light Microscopy

#### Important Concepts

A. Microscope parts. You will be required to know the following parts of the compound light microscope, and their functions: Ocular lenses, objective lenses, arm, base, on/off switch, light intensity knob, fine focus, coarse focus, stage adjustment knobs, lamp, condenser, iris diaphragm, stage, stage clips. The following is a list of the parts of the microscope with the function of each. Label the microscope with the correct parts.

| Microscope part | Function |
|---|---|
| Arm | Support body and lenses. Used to carry microscope. |
| Nosepiece | Carries the objective lenses. |
| Base | Supports the microscope. |
| Stage with clip | To put and secure the slide. |
| Iris Diaphragm Lever | Regulates the amount of light passing through the specimen. |
| Condenser lens | Concentrates light before it passes through the specimen to be viewed. |
| Ocular or eyepiece | Magnifies 10 times. |
| Fine-adjustment Knob | Moves the stage up or down small distances. Allows fine focus of the specimen. |
| Objective lenses (4 lenses) | Magnifies 4, 10, 40, or 100 times. |
| Coarse-adjustment Knob | Moves the stage up or down, should only be used when using the low-powered objective. |

Label the compound microscope.

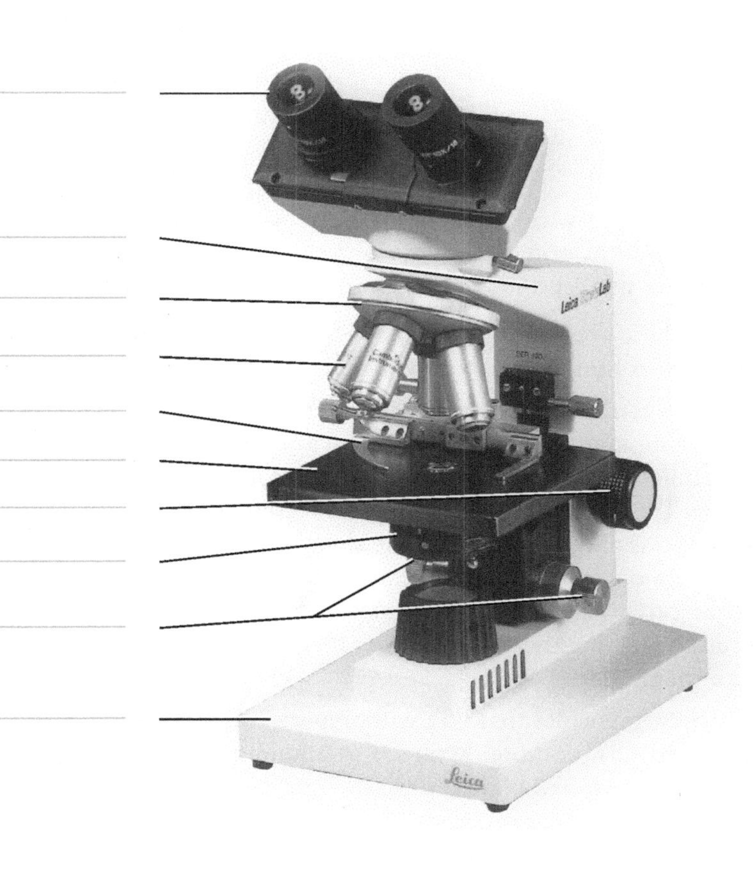

B. Terms to understand.

1. Resolution—clarity of an image.
2. Magnification—how much bigger something appears.
3. Resolving power—how far apart two objects must be in order to see them as separate—shorter wavelength = better resolution.
4. Numerical Aperture—the ability of the lens to capture light from the specimen and make an image—a value that is marked on the lens.
5. Working Distance—the space between the bottom of the objective and the top of the slide.

C. Total Magnification = magnification by ocular × magnification by objective.

D. Field of view—what you can actually see in the microscope.

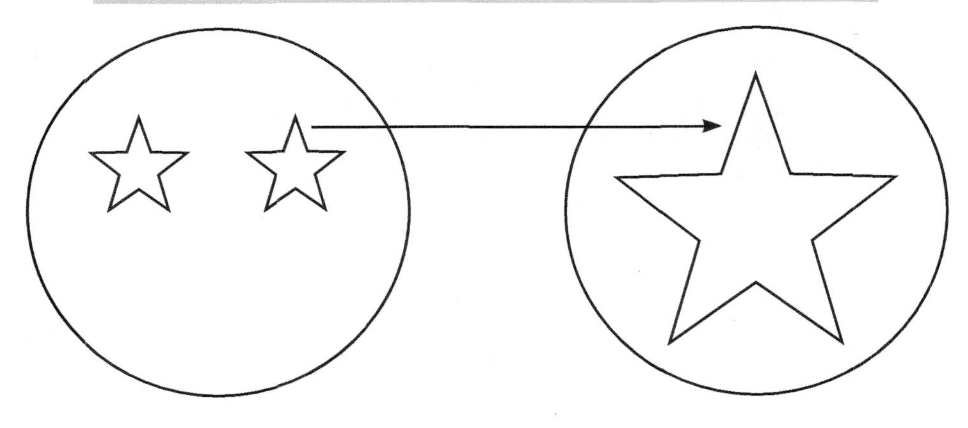

Total magnification = 　　　　　　Total magnification =

E. Depth of Field—refers to how thick the plane of focus is—remember the items we are viewing are 3-dimensional! When using a low magnification, the entire specimen is often in focus. However, as the magnification increases, the depth of field becomes smaller and parts of the specimen may be out of focus while other parts may be in focus.

F. Slide preparation. There are mainly two types of slides that you will be working with this semester: slides that are made for you and slides that you prepare yourself.

- *Prepared slides*—these are factory-made slides that you will use to view a certain item. When asked to view a prepared slide, look carefully at the label to ensure that you are using the correct slide. Then wipe the slide off with a Kimwipe and alcohol, if necessary, to ensure that the slide is clean before use. These slides will be used repeatedly; therefore you should take care to clean them and store them properly when you finish observing them. Always put them back where you got them!
- *Wet mounts*—these are slides that you will prepare yourself. When asked to make a wet mount, you are to obtain a glass microscope slide, clean it and place a drop of your specimen onto the slide. You then cover this with a coverslip. To avoid bubbles, place the coverslip at an angle, and then slowly lower it over the specimen. When you are finished with this slide, please discard the coverslip into the glass disposal box and wash the slide for reuse. If you are finished for the day, please place the slide by the sink.

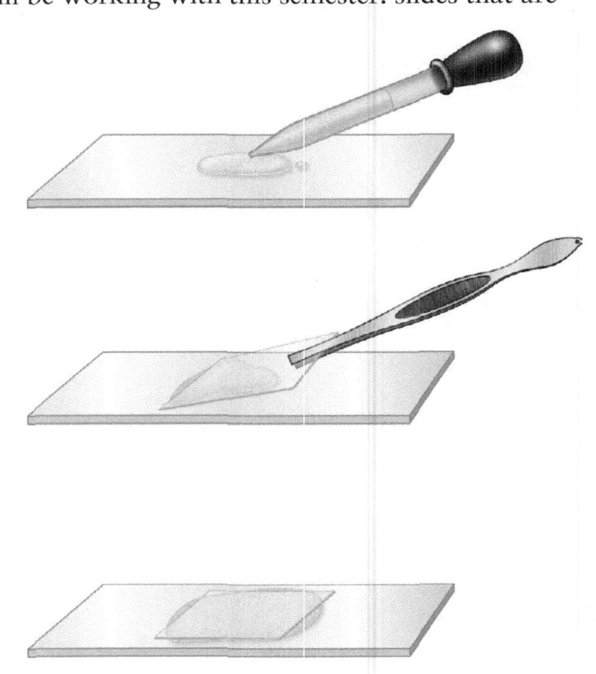

## Activity 2—Focusing your microscope

Before starting our exercises and observing different organisms it is important to keep in mind the following:

> When viewing your image under different objectives, the image remains nearly in focus as you move from one magnification to the next. This is due to the fact that most light microscopes are **parfocal,** which means that when you change the magnification, the image remains nearly in focus. In addition, most microscopes are **parcentral,** meaning that when focused in the center of the field of view, they will remain centered when objectives are switched.

1. Clean all microscope lenses, including the oculars, with lens paper before you begin.
2. Always **start with the low-power objective** (4×)—the scanning objective and the stage all the way down.
3. Use the **coarse-focus knob** to move the stage up until you find an image.
4. **Scan** your slide by using the stage adjustment knobs until you find something interesting.
5. **Center** the object you are interested in.
6. **Increase magnification—change to the next highest power objective.**
7. Use the **fine-focus knob** to bring the image you are interested in into focus.
8. **Increase magnification—change to the next highest power objective.**
9. Use the **fine-focus knob** to bring the image you are interested in into focus.
10. If using the 100× objective, oil can be used. Consult your instructor for proper use of oil.
11. **NEVER**
    a. use the coarse-objective knob when using an objective other than the 4×; this can make it harder to view your specimen, and may break your lenses.
    b. use oil with any objective other than the 100×; this will ruin the lenses. In Biology 101, you will not be using oil; the instructors will be adding it for you.

Observe and draw a prepared slide of a rotifer.

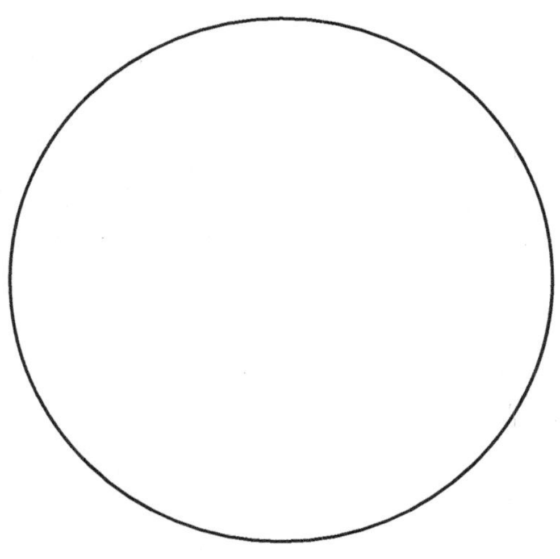

Make a wet mount of the rotifers, then view with the compound light microscope using the 4× and the 10× objectives. **Compare to the rotifers viewed with the dissecting microscope.**

Total magnification = ____________ Total magnification = ____________

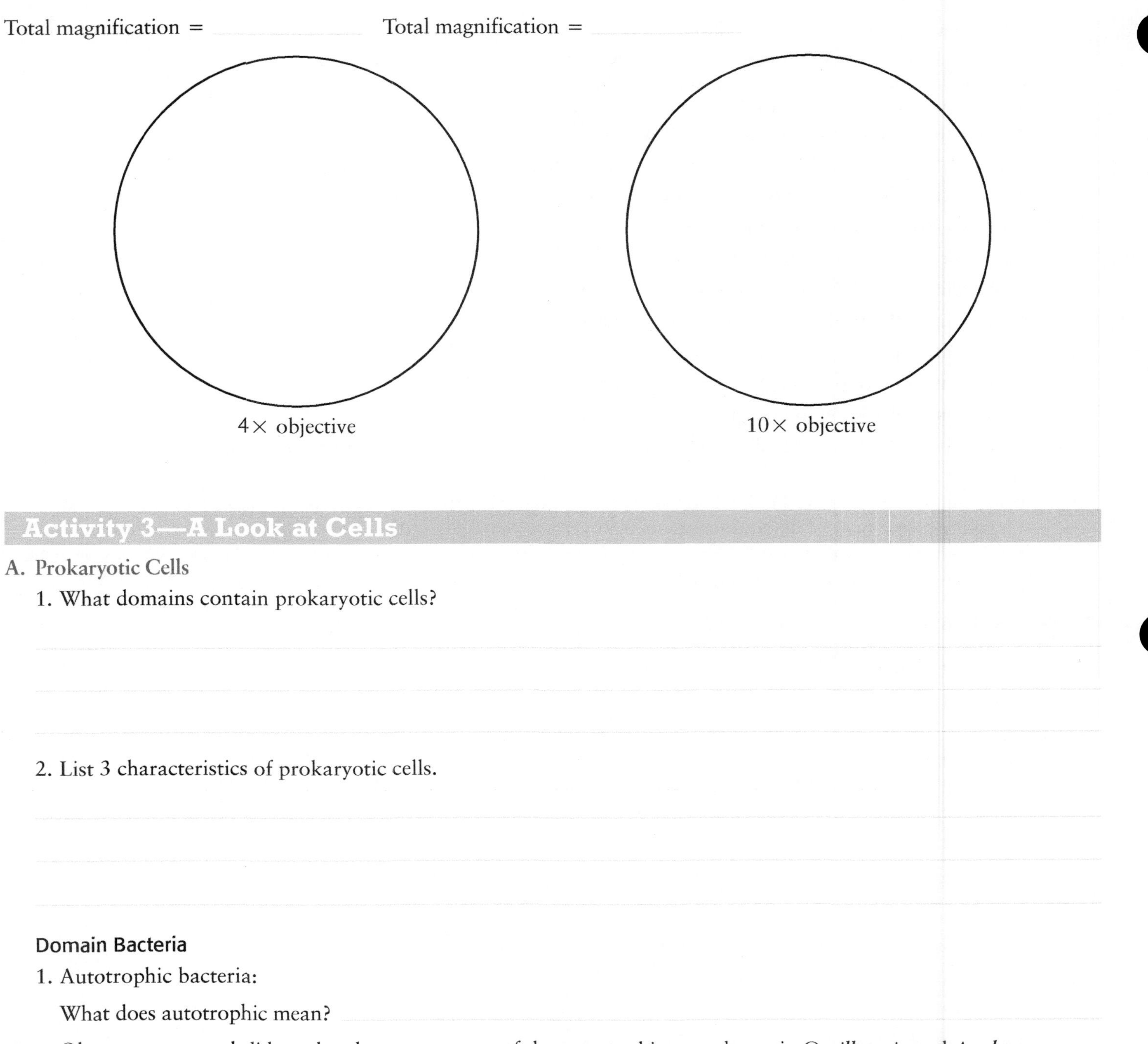


## Activity 3—A Look at Cells

A. Prokaryotic Cells

1. What domains contain prokaryotic cells?

_______________________________________________

_______________________________________________

_______________________________________________

2. List 3 characteristics of prokaryotic cells.

_______________________________________________

_______________________________________________

_______________________________________________

### Domain Bacteria

1. Autotrophic bacteria:

   What does autotrophic mean? _______________________________________________

   Observe a prepared slide and make a wet mount of the autotrophic cyanobacteria *Oscillatoria* and *Anabaena*. View these organisms using the 10× and the 40× objective lenses, then draw the organisms at 40× below. Be sure to LABEL your drawings and make them as accurate as possible—this is what you will use to study. DRAW THE LIVE and FIXED ORGANISMS AT 40×.

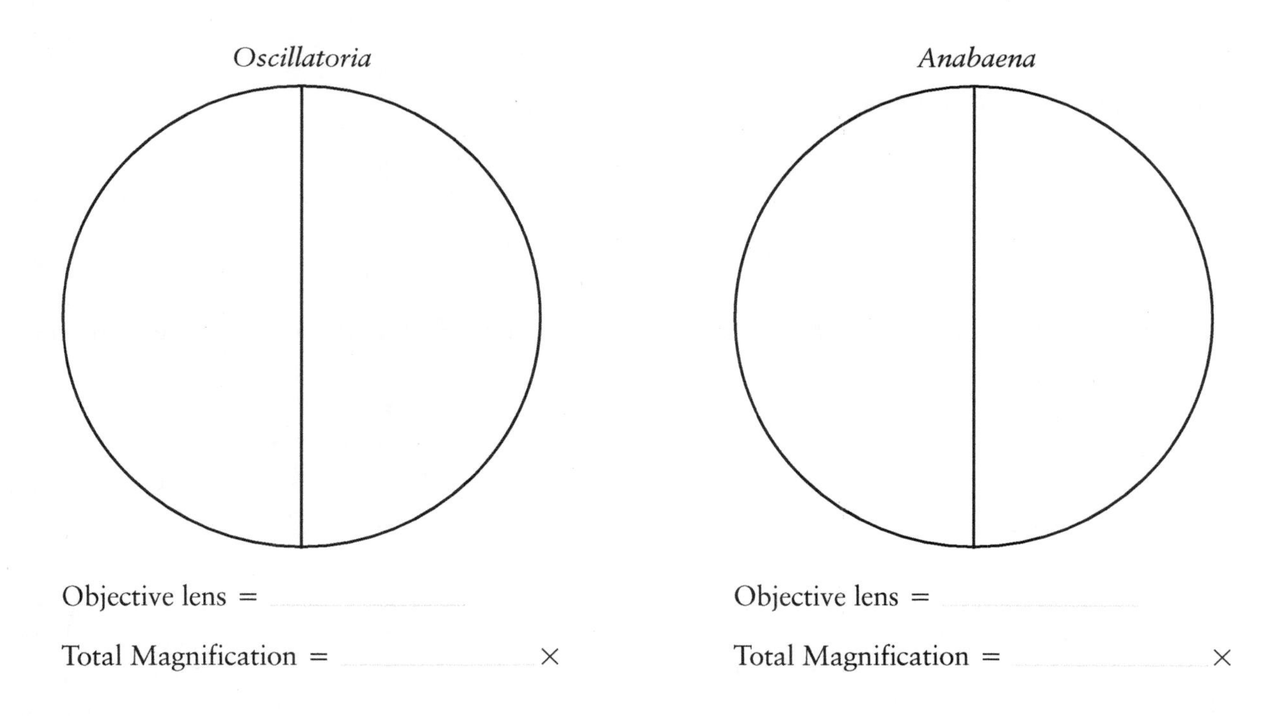


Objective lens =

Total Magnification = ×

Objective lens =

Total Magnification = ×

2. Heterotrophic Bacteria:

What does heterotrophic mean?

Demonstration slides of three primary shapes of heterotrophic bacteria have been set up for you. Please draw each of these shapes.

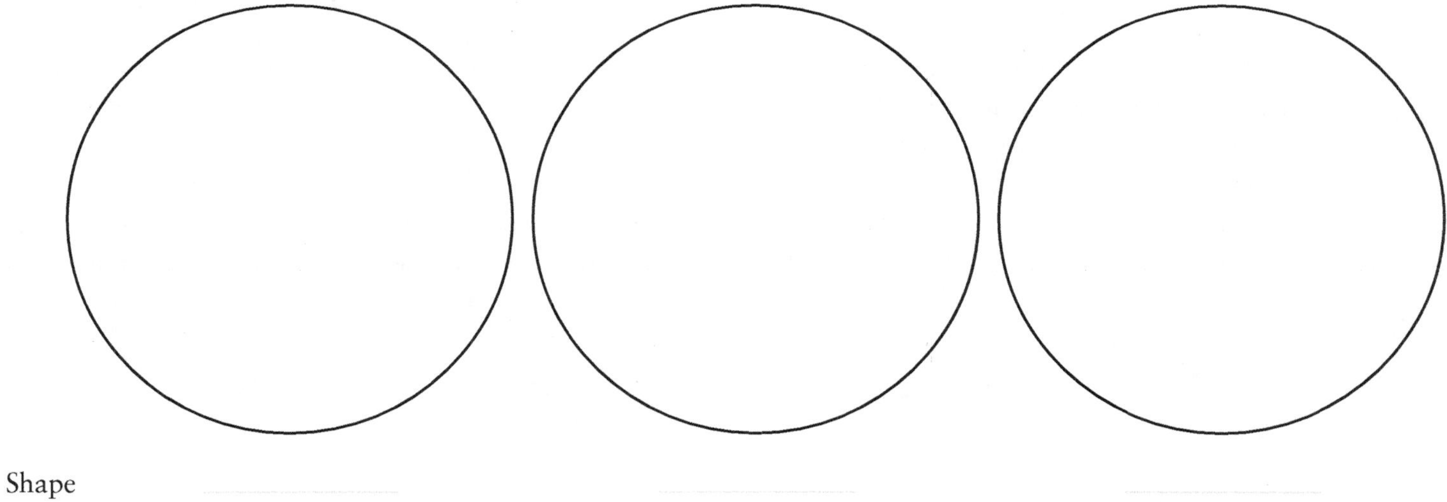

Shape

Description:

Objective used

Total Magnification

**Fun Fact**

*"While you think of yourselves as human, I think of you as 90% bacteria!"*
—Bonnie Bassler

B. Domain Eukarya

List 3 characteristics of the Eukaryotic Cell:

1. **Plant Cells**—Make a wet mount of an *Elodea* leaf. Observe and draw as shown below. **Label cell wall, chloroplasts, central vacuole.**
2. Draw at 40× Draw at 400×

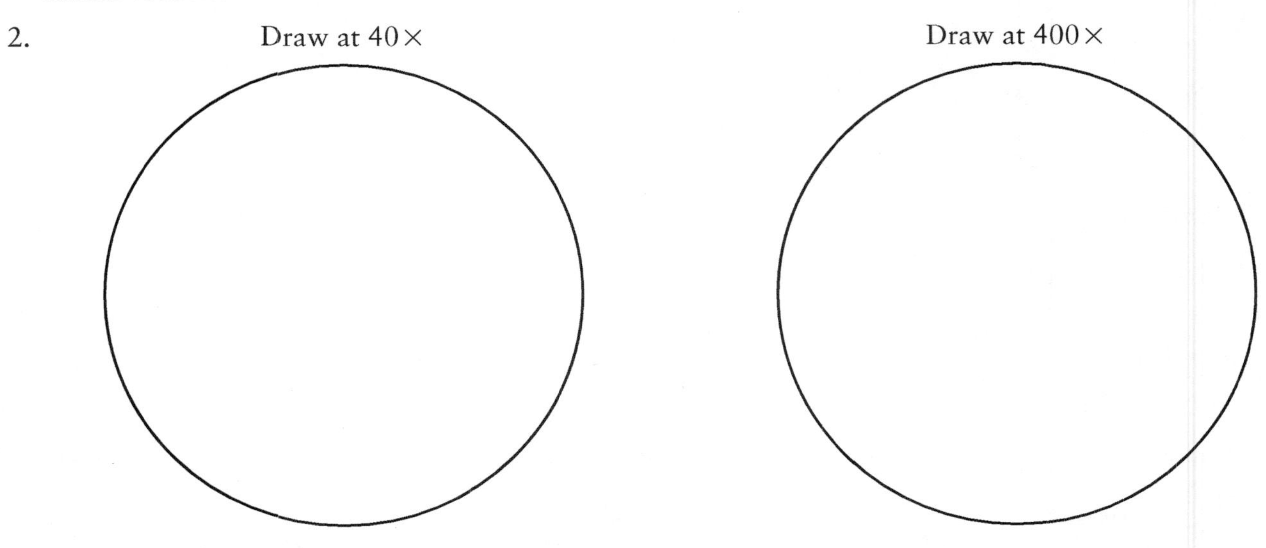

3. **Animal Cells**—Make a wet mount of your cheek cells stained with methlyene blue according to the following directions.
   a. Place a drop of methlyene blue on a clean slide.
   b. Gently scrape the inside of your mouth with the flat side of a toothpick.
   c. Smear the cells you collected onto the dye.
   d. Cover with a cover slip and observe. **Draw and label plasma membrane, nucleus, organelles in cytoplasm.**
   e. **When done, dispose of these slides in the Biohazard Beaker located on the side benches of the lab room.**

Draw at 10× Draw at 40×

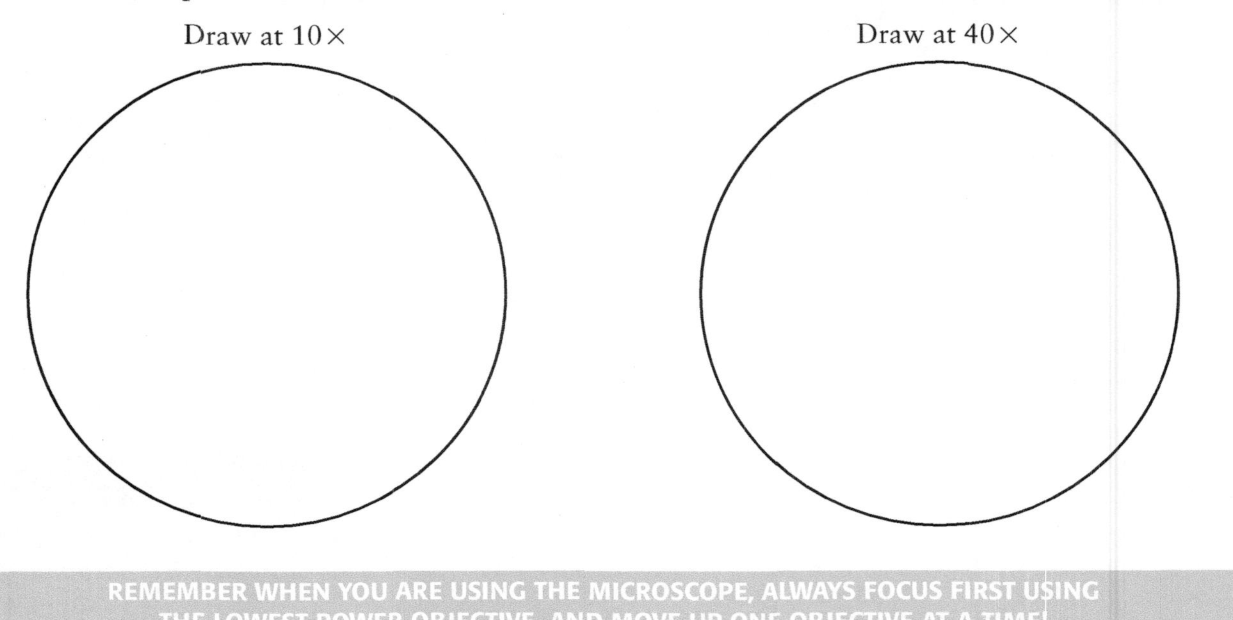

**REMEMBER WHEN YOU ARE USING THE MICROSCOPE, ALWAYS FOCUS FIRST USING THE LOWEST POWER OBJECTIVE, AND MOVE UP ONE OBJECTIVE AT A TIME!**

When you are finished using the microscope, please follow these steps for proper storage.

1. Clean all glass surfaces with lens paper.
2. Remove the slides and put away in the proper location.
   a. If it is a prepared slide, place it back in the rack where you got it.
   b. If it is a wet mount or any slide you made, put the coverslip in the glass disposal box and wash the slide with water. Place in the drying racks near the sink.
3. **Put the stage all the way down and put the low-power objective in place.**
4. Wrap the cord around the microscope.
5. Carry the microscope with two hands to the storage racks.
6. Place the microscopes on the shelf, making sure that they are in the correctly numbered space.

# Review Questions

**1** Why is only the fine-focus knob used after moving from the low-power objective?

**2** Explain *field of view* and what happens to the field of view as magnification increases.

**3** You are observing *Oscillatoria* under the 4× objective and want to increase magnification. When you switch to the 10× objective you cannot see the organism. What has happened and what would you do?

**4** You are viewing rotifers using the 40× objective. What is the total magnification?

**5** How will you store your microscope at the end of lab? Explain these steps in detail!

# A Summary of How to Focus the Microscope

(Use this in future labs as a quick reference.)

1. Clean all microscope lenses, including the oculars, with lens paper before you begin.
2. Always start with the low-power objective (4×)—the scanning objective and the stage all the way up.
3. Use the coarse-focus knob to move the stage up until you find an image.
4. Scan your slide by using the stage adjustment knobs until you find something interesting.
5. Center the object you are interested in.
6. Increase magnification—change to the next highest power objective.
7. Use the fine-focus knob to bring the image you are interested in into focus.
8. Increase magnification—change to the next highest power objective.
9. Use the fine-focus knob to bring the image you are interested in into focus.
10. If using the 100× objective, oil can be used. Consult your instructor for proper use of oil.
11. When finished, clean all glass surfaces with LENS PAPER and store the microscope on low power (4×) with the stage up.

# Biological Molecules

3

***Before coming to lab***

- ☐ Read the lab
- ☐ Answer the pre lab questions

## PRE LAB QUESTIONS

1. What are the four major macromolecules?

2. What are the building blocks of proteins?

3. What are two functions of lipids?

4. What is one safety precaution that should be followed in today's lab?

5. What is Lugol's used to test for?

6. How can you test for lipids if you are at home?

7. What cells will you use to extract DNA?

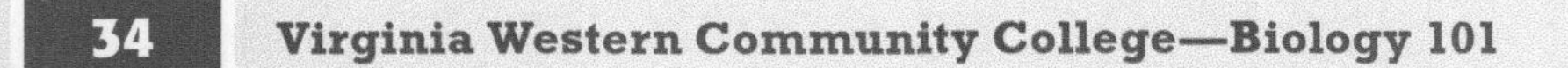

## Purpose of the lab

The purpose of today's lab is to introduce students to the 4 major macromolecules: carbohydrates, proteins, fats and nucleic acids, and their building blocks. In addition students will learn how to use biochemical tests to detect the presence or absence of these macromolecules. Finally, students will learn how to analyze the nutritional content of their diet.

**Dehydration Synthesis**

HO — Short Polymer — H + HO — Monomer — H → $H_2O$

HO — Longer Polymer — H

**Hydrolysis**

HO — Longer Polymer — H + $H_2O$

HO — Short Polymer — H + HO — Monomer — H

## Introduction

All living organisms are composed primarily of organic compounds which are mostly carbon, oxygen, and hydrogen. Large organic molecules are called **macromolecules,** and there are four major groups: carbohydrates, proteins, lipids, and nucleic acids. In this lab we will learn about the first three. The fourth group, nucleic acids, which carry genetic information, will be discussed in a later lab (DNA). With the exception of lipids, these macromolecules (polymers, which are substances composed of a large number of repeating units) are built from small units called monomers by a process called **dehydration synthesis.** A monomer can be released from a polymer by a reverse process called **hydrolysis.**

## Testing for organic compounds

Chemical tests are employed to test for the presence of these macromolecules in a biological sample. For example, a scientist might be interested in determining the contents of a liquid sample found in a bottle. The sample will be identified as the unknown. When testing the unknown sample for a certain compound, the scientist will also test a known sample for the same compound for the sake of comparison (the control). The results from the test of this sample will be compared to results from a control sample. Usually two kinds of controls are tested: a negative control that does not have the compound of interest and therefore should give a negative result and a positive control that has the compound and should show a positive result. There are several tests available to identify the presence of these macromolecules; however, in this lab we will use some of the most common methods to test for carbohydrates, lipids and proteins.

# A. Carbohydrates

Carbohydrates are organic compounds that contain carbon, hydrogen and oxygen in a 1:2:1 ratio. They are sugars, and act as an instant or stored source of energy for the body as well as for structural support. They are classified as: monosaccharides, disaccharides or polysaccharides.

A. **Monosaccharides** are simple sugars such as glucose or fructose.
B. **Disaccharides:** paired monosaccharides such as sucrose and lactose.
C. **Polysaccharides:** chains of three or more monosaccharides such as starch, glycogen or cellulose.

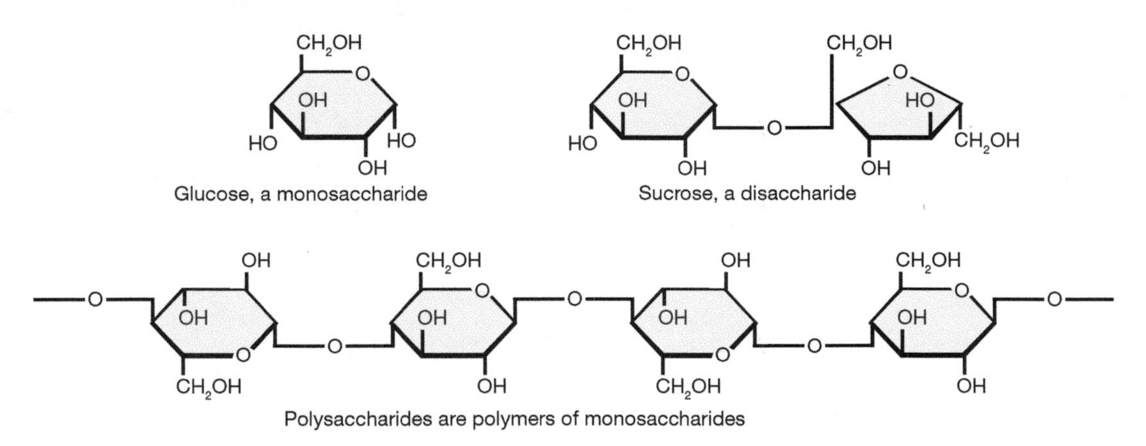


Glucose, a monosaccharide

Sucrose, a disaccharide

Polysaccharides are polymers of monosaccharides

## Testing for Carbohydrates

### A-1 Simple sugars

To test for the presence of simple sugars such as glucose and fructose, **Benedict's solution** is used. Benedict's solution is pale blue and contains blue cupric ions (Cu++). When heated, the cupric ions join with certain sugars and are reduced (gain electrons) to insoluble, red cuprous ions (Cu+). The solution changes from *blue to reddish brown* as more cuprous ions are formed. Only sugars that are able to reduce the cupric ion will produce the color change. Reducing sugars contain a free aldehyde functional group. Glucose is a reducing sugar. Many disaccharides do not react with Benedict's reagent because the reactive groups are often not exposed.

> **Fun Fact**
>
> Benedict's solution was used to test the urine of diabetics before more advanced methods were developed.

## Procedure

1. Half fill a 250 mL beaker with water and put on a hot plate to boil.
2. Obtain 8 test tubes in a rack and label them 1–8.
3. Use a disposable pipette to transfer 2 mL of the correct solution into the tubes according to the following table. **A different pipette should be used for each solution.**

| Tube # | Solution | Expected result |
|---|---|---|
| 1 | Distilled water | |
| 2 | Glucose | |
| 3 | Fructose | |
| 4 | Sucrose | |
| 5 | Lactose | |
| 6 | Starch | |
| 7 | Diet soda | |
| 8 | Sweetened soda | |

4. Make sure your water is boiling before you proceed.
5. Add 2 mL Benedict's solution to each of the tubes.
6. Agitate the tubes.
7. Place all 8 tubes in the boiling water carefully, using tongs, for three minutes.
8. Remove the tubes and record the color and results.
9. Go over results with your instructor before washing tubes.

| Tube # | Solution | Color | Results |
|---|---|---|---|
| 1 | Distilled water | | |
| 2 | Glucose | | |
| 3 | Fructose | | |
| 4 | Sucrose | | |
| 5 | Lactose | | |
| 6 | Starch | | |
| 7 | Diet soda | | |
| 8 | Sweetened soda | | |

**Answer the following questions:**

1. Which of the tubes is the negative control?
2. Which of the tests is the positive control?
3. Explain the test results for starch.
4. Both lactose and sucrose are disaccharides; why did results differ?
5. Explain the results for the diet soda test.

## A-2 Starch

To test for starch, **Lugol's reagent** (iodine) is used. Lugol's distinguishes starch from other carbohydrates. In this test iodine interacts with the coiled molecules that make up starch and changes color from yellow-brown into bluish-black.

What food product do you expect has starch and will test positive if tested?

## Procedure

1. Obtain 7 test tubes in a rack and label them 1 to 7
2. Use a disposable pipette to transfer 3 mL of the correct solution into the tubes according to the following table. **A different pipette should be used for each solution.**

| Tube # | Solution | Expected result |
|---|---|---|
| 1 | Distilled water | |
| 2 | Glucose | |
| 3 | Starch | |
| 4 | Potato juice | |
| 5 | Cereal | |
| 6 | Soda | |
| 7 | Diet soda | |

3. Add 9 drops of Lugol's reagent to each tube.
4. Record color and results in the following table.
5. Go over the results with your instructor before you wash the tubes.

| Tube # | Solution | Color | Results |
|---|---|---|---|
| 1 | Distilled water | | |
| 2 | Glucose | | |
| 3 | Starch | | |
| 4 | Potato juice | | |
| 5 | Cereal | | |
| 6 | Soda | | |
| 7 | Diet soda | | |

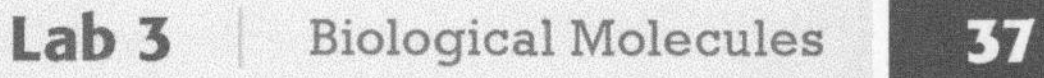

**Answer the following questions:**

1. Which of the above tubes is the negative control?
2. Which of the above tubes is the positive control?
3. Why can't we use Benedict's to test for starch if it is made of glucose?
4. If you add Lugol's to a solution in a test tube and it turns orange-brown, what would you conclude and how would you interpret the results?

> **Fun Fact**
>
> A money-tester pen is used to identify counterfeit money. The ink in the pen is mixed with iodine and changes to black when it reacts with starch. Real money is treated and all traces of starch are removed. Counterfeiters have not mastered this technique and therefore the paper has starch and ink will turn blue-black on this paper.

# B. Lipids

Lipids are organic compounds that are hydrophobic and therefore insoluble in water but soluble (dissolve) in organic solvents like ether and acetone. Lipids are composed mostly of carbon and hydrogen and very little oxygen. The following drawing illustrates saturated and unsaturated fats. What are the key differences between these?

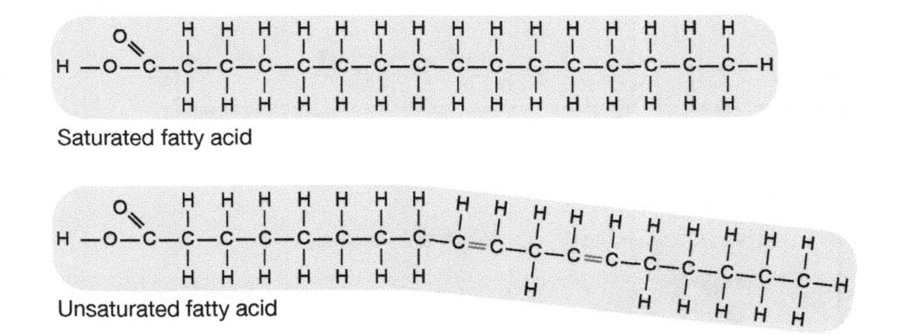


Lipids serve several functions in the body, mainly as an abundant source of energy or structural as part of cell membranes. Examples of lipids include phospholipids, waxes, steroids and fats.

## Testing for lipids

### B-1 Paper towel method

Without knowing it, you have probably done and read the results of this test before. This is a simple test used to identify fats. Have you ever noticed a stain of grease on a paper bag that has French fries in it? Fats usually leave a stain on a paper. Predict whether lipids are present in the following compounds:

| Sample | Solution | Expected result |
|---|---|---|
| 1 | Distilled water | |
| 2 | Vegetable oil | |
| 3 | Chicken broth | |
| 4 | Egg white | |
| 5 | Fat-free salad dressing | |
| 6 | Salad dressing | |

## Procedure:

1. Obtain a paper towel from the lab and cut it into 6 pieces.
2. Label the pieces 1 to 6.
3. Put a drop of solution on each of the paper pieces according to the table below. **A different pipette should be used for each solution.**
4. Let it dry for 15 minutes; while this dries proceed to Sudan IV test.
5. Read result; if lipid is present a stain would be visible.
6. Fill in the table with your results.

| Sample | Solution | Observations | Results |
|---|---|---|---|
| 1 | Distilled water | | |
| 2 | Vegetable oil | | |
| 3 | Chicken broth | | |
| 4 | Egg white | | |
| 5 | Fat-free salad dressing | | |
| 6 | Salad dressing | | |

7. Consult with your instructor before disposing of paper.

**Answer the following questions:**

1. Which of the above was the negative control? ______
2. Which of the above was the positive control? ______
3. Which of the products tested has the highest amount of lipids? ______

### B-2 Sudan IV test

Sudan red is a chemical that is insoluble in water but soluble in lipids. To test for lipids, Sudan IV is added to the solution, which will interact with any lipid present and turn it bright red. Observe water test for negative control. Predict whether lipids are present in the following compounds:

| Section # | Solution | Expected result |
|---|---|---|
| 1 | Distilled water | |
| 2 | Vegetable oil | |
| 3 | Chicken broth | |
| 4 | Fat-free salad dressing | |
| 5 | Egg white | |
| 6 | Egg yolk | |
| 7 | Salad dressing | |

## Procedure

1. Obtain 7 test tubes and label them 1 to 7.
2. Add 3 mL of distilled water to each tube.
3. Add 3 mL of the test solution to each tube according to the following table. **A different pipette should be used for each solution.**
4. Add 9 drops of Sudan IV into each tube.
5. Agitate by shaking side to side.
6. Add 2 mL distilled water to each tube—DO NOT MIX.
7. Record results in the following table.

| Tube # | Solution | Observations | Results |
|---|---|---|---|
| 1 | Distilled water | | |
| 2 | Vegetable oil | | |
| 3 | Chicken broth | | |
| 4 | Fat-free salad dressing | | |
| 5 | Egg white | | |
| 6 | Egg yolk | | |
| 7 | Salad dressing | | |

8. Go over the results with your instructor before washing tubes.

**Answer the following questions:**

1. Explain the results of the vegetable oil tube
2. Which test is more sensitive to small amounts of lipids—paper or Sudan IV?
3. What are limitations of these tests?

## Disturbing Fact

325,000 people die yearly from obesity.

## C. Proteins

Proteins are organic compounds that are very abundant in living organisms and have multiple functions. They provide support, defense, movement, storage, and regulation. Enzymes are proteins and are essential for reactions to occur at a fast rate within cells. Some examples of proteins include: actin (movement, in cytoskeleton and muscles), hemoglobin (carry oxygen), antibodies (defense), lactase (break lactose), and keratin (found in skin, nails and hair). Proteins are polymers of amino acids. There are 20 different amino acids that make up all proteins. Each amino acid has a carboxyl group and amino group. Amino acids combine by covalent bonds between the amino group of one amino acid and the carboxyl group of another amino acid. These bonds are called peptide bonds.

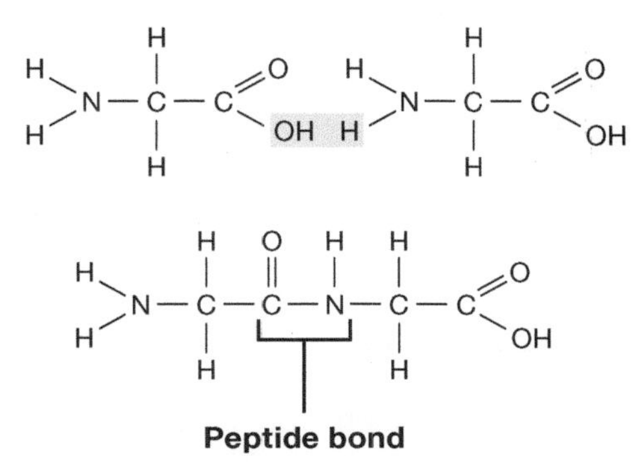


A molecule of water is removed from two glycine amino acids to form a peptide bond.

### Testing for Proteins

To test for proteins, **Biuret reagent** is used. Biuret is 1% solution of $CuSO_4$ and is blue in color. In this test, $Cu^{+2}$ interacts with the peptide bond and forms a violet color in which the intensity of the color is directly related to the number of peptide bonds. This means the more peptide bonds present the darker the violet color will be. However, Cu has to complex with at least 4 or 6 peptide bonds for the color to change. Therefore, very short protein chains and free amino acids will not give a positive result with Biuret. Predict whether protein is present in the following compounds:

| Tube # | Solution | Expected result |
|---|---|---|
| 1 | Distilled water | |
| 2 | Vegetable oil | |
| 3 | Glucose | |
| 4 | Egg white | |
| 5 | Egg yolk | |
| 6 | Amino acid solution | |
| 7 | Albumin solution | |

## Procedure

1. Obtain 7 test tubes and label them 1 to 7.
2. Pipette 1 mL of the test solutions into each tube as described in the table. **A different pipette should be used for each solution.**
3. Add 10 drops of Biuret solution to each tube.
4. Agitate by mixing side to side.
5. Wait 2 minutes.
6. Read the results; record in the table.

| Tube # | Solution | Color | Result |
|---|---|---|---|
| 1 | Distilled water | | |
| 2 | Vegetable oil | | |
| 3 | Glucose | | |
| 4 | Egg white | | |
| 5 | Egg yolk | | |
| 6 | Amino acid solution | | |
| 7 | Albumin solution | | |

7. Go over the results with your instructor before washing the tubes.

**Answer the following questions:**

1. Which of the above substances are proteins?
2. Which of the above solutions is the positive control?
3. Do free amino acids result in a positive Biuret test? Explain your results.
4. Which of the solutions tested has the highest amount of proteins?

### Fun Fact

Do you ever wonder why shrimp turn pink when cooked?

There are several pigments in the shell of shrimp. *Astaxanthin* is a pink or red pigment that is wrapped up in dark protein chains. Whenever you cook the shrimp, the proteins denature and uncoil exposing the red pigment and the shell changes color.

## D. Nucleic Acids

### Isolation of YOUR own DNA!

During the final activity today, you will be allowed to isolate your own DNA and make a necklace from it! Your instructor will provide a handout that details this very easy procedure. Basically you are going to collect some of your cheek cells and take your DNA from them. Remember that the DNA in eukaryotic cells is located in the nucleus, so both the cell and the nucleus must be opened to get the DNA out. To open the cell and the nucleus, both of which are surrounded by lipid bilayer membranes, a detergent will be used. Then the DNA will be collected using cold ethanol. As you are collecting your DNA, think about the composition of DNA (A, T, G, C) and the fact that this material is what separates so accurately during every mitotic division!

# Review Questions

1. Fill in the following table.

| Macromolecule | Building block | Example | Reagent to test | Positive outcome |
|---|---|---|---|---|
| | | | | |
| | | | | |
| | | | | |
| | | | | |
| | | | | |
| | | | | |

# Diffusion and Osmosis

4

***Before coming to lab***

- ☐ Read the lab
- ☐ Answer the pre lab questions

## PRE LAB QUESTIONS

1. What is the difference between osmosis and diffusion?

2. What is a solvent? Give an example.

3. What is selective permeability?

4. What is the plant cell that we will use in today's lab?

5. What is the animal cell that we will observe in today's lab?

6. What is the cell membrane made of?

## Purpose of the lab

The purpose of today's lab is to introduce students to the selectively-permeable cell membrane by observing and studying diffusion of molecules and osmosis of water through these membranes under varying conditions. In addition, students will explore how plant and animal cells behave in different osmotic environments.

## Introduction

All living things have certain requirements that they need to stay alive. One important example is that they must be able to exchange gases, nutrients and wastes with their environment. This exchange occurs at the cellular level and since all cells are surrounded by a cell membrane, these molecules must pass through that membrane. The cell membrane is made of a phospholipid bilayer, which makes it a highly **selective membrane**, meaning it allows certain molecules to cross while preventing others. Substances entering and leaving the cell are usually dissolved in water (the universal **solvent**), which is essential for life. The substances that are dissolved in the water are referred to as **solute,** and the mixture of solvent and solute is the **solution.**

If you mix salt and water in a cup:

What is the solvent? ____________

What is the solute? ____________

What is the solution? ____________

Materials can cross the cell membrane by two different mechanisms: active transport and passive transport. **Active transport** requires expenditure of energy and moves molecules from an area of low concentration to an area of high concentration (against the concentration gradient). **Passive transport** requires no energy and moves molecules from high concentration to that of low concentration (down the concentration gradient). The main process of passive transport is **diffusion,** which refers to the movement of dissolved molecules across a semipermeable membrane from an area of high concentration to lower concentration. One form of diffusion is referred to as **facilitated transport,** which requires transport proteins to facilitate this transport. **Osmosis** is a special form of diffusion that involves water where water moves across a semipermeable membrane from an area with high water concentration to that with low water concentration.

When a membrane separates two areas that differ in the concentration of a solute, osmosis will occur and water will move from the **hypotonic** solution (lower solute concentration) into the **hypertonic** solution (high solute concentration). Water will move down its concentration gradient. This will continue until both sides have the same solute concentration, at which time **equilibrium** is said to be reached. If the concentration of the solute is the same on both sides of the semipermeable membrane, the solutions are said to be isotonic and there will be no net movement of water. Many cells exist in **isotonic** solutions in which the concentration of solutes is the same inside and outside the cells.

## Practicing the concepts

### Exercise 1

Assume that a semipermeable balloon with a solution A is placed in a beaker within solution B. For each of the different scenarios below:

1. Use an arrow to show where water will go.
2. Indicate which solution is hypertonic, hypotonic or isotonic.
3. Predict what will happen to the balloon.

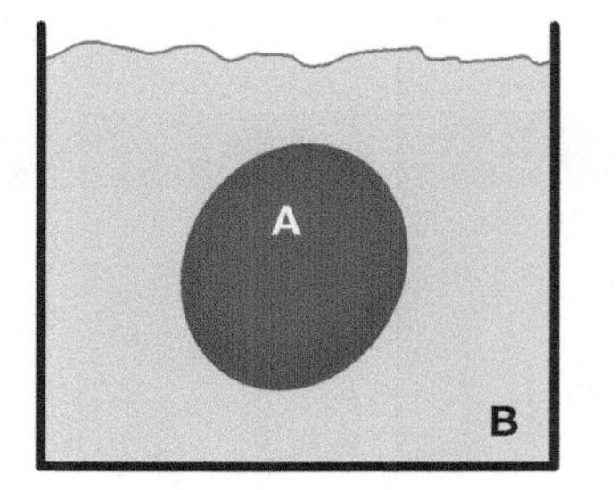


Scenario 1: Solution A is 1% salt; Solution B is 5% salt.

1. 
2. 
3. 

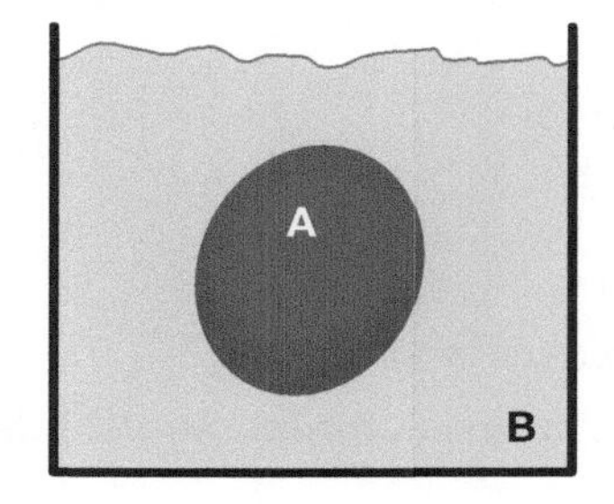


Scenario 2: Solution A is 10% salt; Solution B is 5% salt.

1. 
2. 
3. 

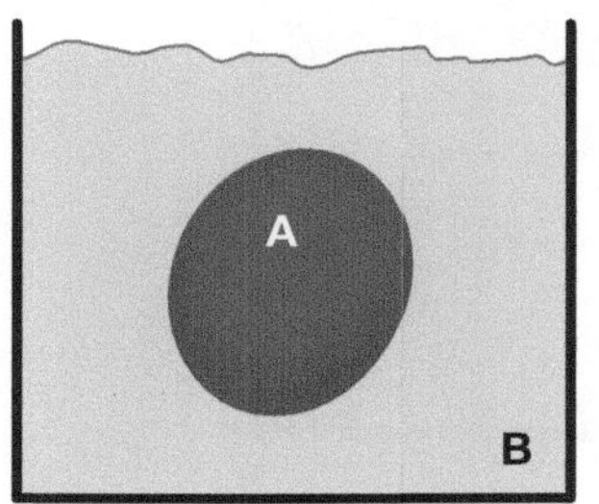


Scenario 3: Solution A is 1% salt; Solution B is 99% water.

1. 
2. 
3. 

### Exercise 2

The cells of a fish that lives in fresh water have more solutes than the surrounding solution. Draw a picture of the fish and explain where water will go. Try to guess how a fish can survive in such an environment.

## Activity 1—Diffusion across a semipermeable membrane—the effect of molecular weight

Molecular weight of a molecule is very important in diffusion. Sometimes smaller molecules can pass through a membrane while larger ones cannot. In this experiment we will investigate how molecular weight can influence diffusion, by using dialysis tubing. This tubing is made out of synthetic material that has microscopic holes in it. Your instructor will demonstrate how to use dialysis tubing to make bags that can be filled with solutions. For a substance to diffuse through this membrane, it has to pass through the microscopic holes. We will investigate if starch and/or chloride ions can diffuse across the membrane. Why do you think we are using these two solutes? What is the difference and why were they picked?

## Procedure

1. Obtain a section of precut dialysis tubing.
2. Clamp one of the ends.
3. Add 10 mL starch solution into the **bag** and clamp the other end.
4. Pour 250 mL NaCl solution into a beaker.
5. Rinse bag with distilled water before proceeding.
6. Immerse the bag in the **beaker** solution.
7. Record the time.
8. Wait 60 minutes.
9. Obtain 4 test tubes and number them 1 to 4.
10. After 60 minutes empty the contents of the bag into a clean beaker labeled "Bag."
11. Transfer 2 mL of each of the bag and beaker solutions into the tubes, **according to the following table,** to test for starch and chloride ions.

| Tube # | Solution | Test for |
|---|---|---|
| 1 | Bag | Starch |
| 2 | Bag | Chloride ion |
| 3 | Beaker | Starch |
| 4 | Beaker | Chloride ion |

12. Add 9 drops of Lugol's to test for starch.
13. Add 4 drops of 2% silver nitrate ($AgNO_3$), which will precipitate as white silver chloride ($AgCl$) if chloride is present.
14. Record your findings in the table.

| Tube # | Solution | Test for | Observations | Conclusions |
|---|---|---|---|---|
| 1 | Bag | Starch | | |
| 2 | Bag | Chloride ion | | |
| 3 | Beaker | Starch | | |
| 4 | Beaker | Chloride ion | | |

15. Explain your findings:

## Activity 2—Osmosis

During this activity, students will use potatoes to understand the process of osmosis.

**Materials:**

- ❑ 3 small beakers (150 or 250 mL)
- ❑ 3 two cubic centimeters ($cm^3$) pieces of potato
- ❑ Solutions (listed in table)

| Solution | Mass before soaking (grams) | Mass after soaking (grams) | Mass Difference (after−before) | % change |
|---|---|---|---|---|
| Distilled $H_2O$ | | | | |
| 10% sucrose | | | | |
| 50% sucrose | | | | |

## Procedure

1. Cut the potato into three roughly 2 $cm^3$ pieces.
2. Weigh each piece separately and record the mass before soaking.
3. Place each piece of potato into a separate beaker and label each beaker with one of the solutions listed above.
4. Pour enough of the solution over each potato piece to completely submerge it.
5. Allow to soak for one hour.
6. After one hour remove each potato piece, lightly dry them off, and measure their mass.
7. Record the mass in the table and then calculate the % change in mass using the following equation:

$$\frac{\text{Mass after soaking} - \text{mass before soaking}}{\text{Mass before soaking}} \times 100 = \%\ \text{change in mass}$$

8. Explain your findings, and identify the *hypertonic*, *isotonic*, and *hypotonic* solutions.

In this experiment, what is your independent variable?

What is your dependent variable?

What variables were constant?

## Activity 3—Osmosis and living cells

In this experiment we will investigate osmosis in both plant and animal cells.

Mammalian red blood cells (RBC) are enucleated, and the serum is isotonic with the cytoplasm concentration, which makes the RBC assume a biconcave (doughnut-like) shape. However, if a RBC is placed in either hypotonic or hypertonic solution, water will either go in or out of the cell, which results in a different shape. Below is a demonstration of the effect of different solutions on RBC.

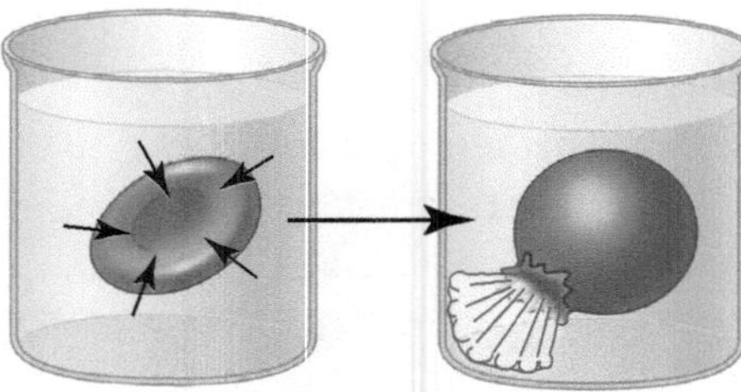

Hypotonic solution

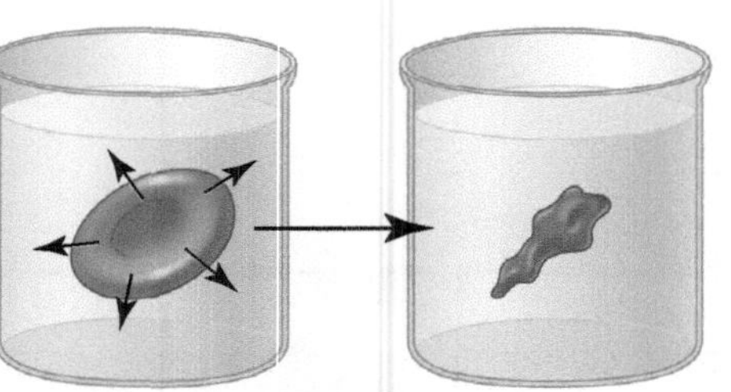

Hypertonic solution

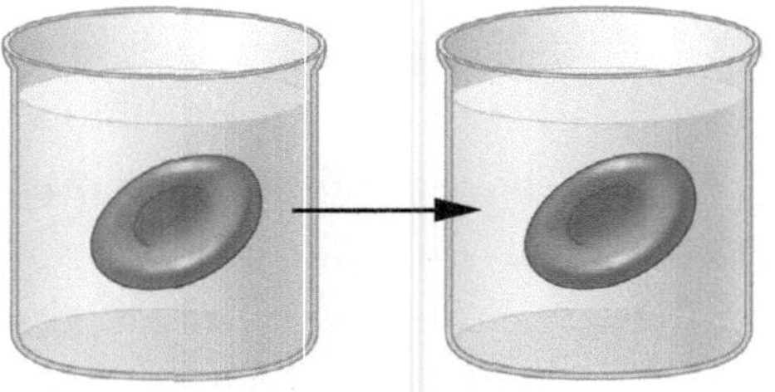

Isotonic solution

Applying the concept:

1. If a red blood cell is placed in a hypertonic solution:
   a. Where will the water go?

   b. What will happen to the cell?

2. If a red blood cell is placed in a hypotonic solution:
   a. Where will the water go?

   b. What will happen to the cell?

## Procedure for animal cells

1 Set up three microscopes on your table.
2 Obtain three slides.
3 Number them in the corner 1, 2 and 3.
4 Make wet mounts of sheep red blood cells as described below.
   a. Slide 1: 1 drop blood with 1 drop 0.9% saline.
   b. Slide 2: 1 drop blood with 1 drop 5% saline.
   c. Slide 3: 1 drop blood with 1 drop distilled water.
5 Observe the slides and compare to each other.

| | 0.9% Nacl | 5% Nacl | distilled water |
|---|---|---|---|
| Sketch | | | |
| Explain | | | |

**6** Answer the following questions:

a. Which solution is hypertonic to red blood cells?

b. Which solution is hypotonic to red blood cells?

c. Which solution is isotonic to red blood cells?

Plant cells are enclosed by a rigid cell wall that prevents it from bursting when water rushes in.

1. If a plant cell is placed in a hypertonic solution:

   a. Where will the water go?

   b. What will happen to the cell?

2. If a plant cell is placed in a hypotonic solution:

   a. Where will the water go?

   b. What will happen to the cell?

## Procedure for plant cells

**1** Obtain three slides.

**2** Number them in the corner 1, 2 and 3.

**3** Make wet mounts of *Elodea* as described below.

a. Slide 1: 1 leaf of *Elodea* plant with 1 drop pond water in it.

b. Slide 2: 1 leaf of *Elodea* plant with 1 drop 5% saline.

c. Slide 3: 1 leaf of *Elodea* plant with 1 drop distilled water.

**4** Observe the slides and compare to each other. Draw and label cell wall, chloroplasts, plasma membrane, and central vacuole, if visible.

| | Pond water | 5% Nacl | distilled water |
|---|---|---|---|
| **Sketch** | | | |
| **Explain** | | | |

**5** Answer the following questions:

a. Which solution is hypertonic to *Elodea* cells?

b. Which solution is hypotonic to *Elodea* cells?

c. Which solution is isotonic to *Elodea* cells?

# Review Questions

**1** When a plant cell is placed in distilled water, what structure prevents it from bursting?

**2** Why would a produce manager at a grocery store spray the vegetables with water several times during the day?

**3** A gardener is having problems with slugs destroying his vegetables. A neighbor suggested adding salt on the slugs to control this problem. Explain whether you think this could solve the problem, and how?

**4** Tom ran out of saline lens solution, so he just used tap water. Do you think this is a good idea? Explain.

**5** Explain why fresh water fish cannot live in the ocean and salt water fish cannot live in fresh water.

# Lab Practical

- Lab Safety
- Scientific method
- Measurement
- Microscopy and Cells
- Biological Molecules
- Diffusion and Osmosis

Don't cram the last night.
Study ahead of time and be ready for the practical

# Enzymes Part 1

***Before coming to lab***

- ☐ Read the lab
- ☐ Answer the pre lab questions

## PRE LAB QUESTIONS

1. What is an enzyme?

2. Define substrate.

3. What type of macromolecule are most enzymes?

4. What are the four levels of protein structure?

5. What is activation energy?

6. What color is ONP?

7. What is a serial dilution?

8. What are two environmental factors that can influence enzyme activity?

9. What causes lactose intolerance?

## Purpose of the lab

The purpose of this lab is to introduce the students to the basics of enzymatic reactions that occur within every living organism. This lab is divided into two weeks to allow for a better understanding of enzymes and how they work. The first week the student will learn how standard curves can be used to estimate enzyme activity and how to properly conduct enzymatic assays. During the second week the student will examine what environmental factors influence enzyme activity.

## Introduction

Enzymes are molecules present in each and every living cell that allow vital chemical reactions to occur within that cell. They are biological catalysts that are involved in almost all metabolic reactions in a cell. *Most* enzymes are proteins, and are therefore subject to the same rules that govern how proteins are put together.

> **Fun Fact**
>
> While most enzymes are proteins, it was discovered in the 1980s that RNA molecules also could fold and function as enzymes. These enzymes are often called ribozymes, and this information indicates that nucleic acids, in particular RNA, were probably the first biologically active molecules on earth.

Proteins are chains of amino acids held together by peptide bonds. As you have already learned, these chains of amino acids fold onto each other, based on the interaction among the various amino acids. This results in a molecule with a very precise three-dimensional shape! It is very important to remember that enzymes are not flat shapes that we draw on paper; they are three-dimensional objects with a predetermined conformation, or shape. Review the 4 levels of protein structure from the material on macromolecules, and describe each level of protein structure below.

Primary structure: ______

Secondary structure: ______

Tertiary structure: ______

Quaternary structure: ______

Where do enzymes come from? Because most enzymes are proteins, they are coded for by regions of the DNA called genes. All the enzymes that a cell produces are made from the DNA of that organism.

What do enzymes do? As previously stated, enzymes allow chemistry to occur in a cell. They accomplish this by lowering the activation energy, which is the energy required to break chemical bonds and allow two molecules to interact. The activation energy is often too great to allow the reaction to occur as quickly as the cell needs it. The enzymes decrease this barrier and allow the cell to function. Enzymes are usually present in the cell in small amounts, and they can be regulated, depending upon the needs of the cell.

How does an enzyme work? Enzymes work by binding to another molecule, called the **substrate**, at a particular place on the enzyme, called the **active site**. The active site is present because of the folding of the molecule—enzymes will not work unless they are in the correct conformation. Further, enzymes are not used up in a chemical reaction—they are recycled and can be used to catalyze the same reaction again. Here is a diagram of a typical enzymatic reaction.

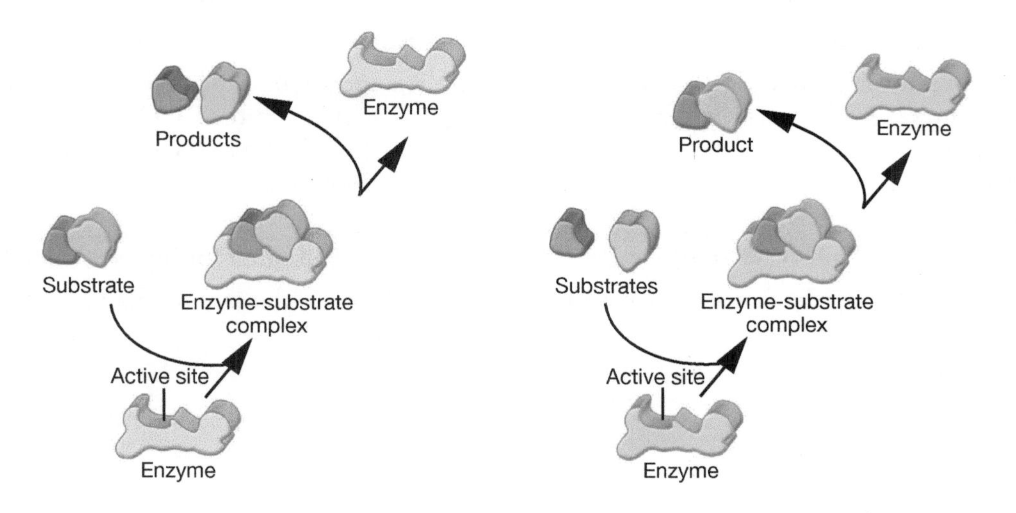

Another concept to emphasize is that enzymes are SPECIFIC. That means that they will usually act on only one substrate. This is because the substrate must fit into the active site, and if the active site is not the right shape, the substrate will not fit and the reaction will not occur. Enzymes are also affected by the physical and chemical conditions of their environment. If anything changes the shape of the enzyme, it will affect its ability to function. Physical conditions like temperature and pH, for example, are critical to proper enzyme function.

Today, you will be using some familiar materials and some unfamiliar chemicals to learn about enzyme activity. For these experiments, the enzyme we will be using is called lactase. As an aside, remember that many enzyme names end in -ase— so that can help you find your enzyme! Lactase is a member of the B-galactosidase family of enzymes, and they are discussed frequently in the scientific literature. In the first set of experiments, we are going to use lactase as the enzyme and a molecule called ortho-nitrophenyl-galactoside (ONPG) as the substrate. (Normally the sugar lactose is the substrate for this enzyme— we will get to that later.) The product of this chemical reaction is ortho-nitrophenolate (ONP). The reason we use this substrate is because it produces a product that has a yellow color, and is therefore very easy to detect and quantify.

The summary equation for the reaction we will be using is as follows:

$$\text{ONPG} + H_2O \xrightarrow{\text{lactase}} \text{ONP} + \text{Galactose}$$

Please label the **enzyme, substrate** and **product** in the above equation. Which of these molecules is yellow?

## Activity 1—Making a standard curve

The product you will be making in today's lab is o-Nitrophenolate (ONP), which is a yellow-colored compound. The first step in the experiment will be to determine how the **color** intensity of the product relates to the **concentration** of the product. Put quite simply, the more yellow your ending solution is, the more ONP is present. To quantify this relationship, you will generate a **standard curve.** A standard curve is a line graph that is generated using KNOWN values and can be used to determine unknown concentrations. The concept of a standard curve is very common practice in research laboratories, and allows scientists to indirectly measure enzyme activity present in an unknown solution. Today, you will generate a standard curve using known quantities of ONP and determining the absorbance of each solution. You will then generate a line graph to show the relationship between the concentration of ONP and absorbance. You will work at your desk as a group of 4 for this experiment. The machine that you will be using is called a spectrophotometer (Spectronic 20), and this machine measures the amount of light that is absorbed by or transmitted through a substance. You will be using this machine in other experiments, so pay close attention!

1. Turn on the Spectronic 20 (Spec 20) instrument and allow it to warm up for 10 minutes.
2. Obtain the **stock solutions** of ONP that have been prepared for you. These concentrations are: 0, 25, 50, 75, 100, 150, 200 and 250 μmoles (μmoles is a unit that measures the amount of a substance present).
3. With the lid to the sample chamber closed, **the digital display should read a wavelength in nm (nanometers) and give a number for A (= absorbance).** If the data are not provided as A, toggle through the A/T/C key until they are.
4. Use the ▲ nm or ▼ nm to **set the wavelength to 420 nm**—this represents the optimal absorbance for ONP.
5. Prepare a blank of phosphate buffer (the material that the ONP is prepared in) in a cuvette and carefully wipe the BLANK with a Kimwipe (to remove any dirt or oils from fingerprints). **Insert the BLANK into the sample chamber,** being sure to align the white mark with the front of the chamber. Close the lid to the sample chamber and **press the 0 ABS/100% T key.** This will calibrate the machine so the reported absorbance is due only to the ONP, not the solvent in which it is dissolved.
6. When the digital display reads 0 (= zero) for absorbance, the spec is calibrated for this wavelength.
7. **Insert the cuvette containing the first stock solution (25 microMolar)** into the sample chamber. Without touching any keys on the spec, read and record the value for A in your data table.
8. Take that cuvette out and repeat this procedure for each stock solution.

9. Record your data in the following table.

| Concentration of ONP (μMol) | Absorbance |
|---|---|
| 0 | |
| 25 | |
| 50 | |
| 75 | |
| 100 | |
| 150 | |
| 200 | |
| 250 | |

Now, use these data to plot a standard curve. Remember to title your graph, and properly label the x and y-axes and use a pencil to draw your graph (so you can neatly erase any errors!).

How will your x-axis be labeled? ____________________

How will your y-axis be labeled? ____________________

**Title:** ____________________

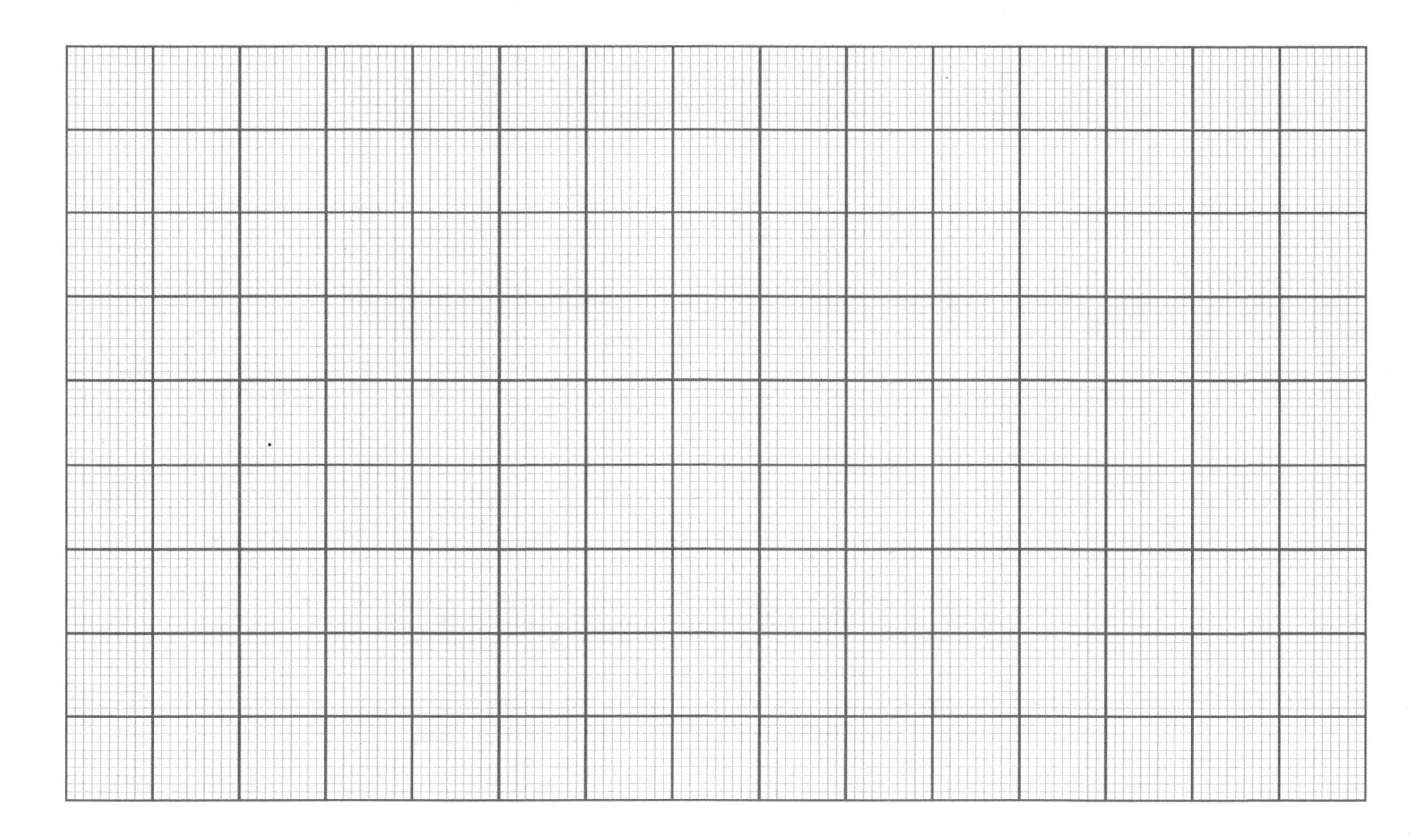

**This will be your standard curve for the remaining experiments today, and you will use this standard curve to determine the amount of ONP produced by an enzymatic reaction.** You will look at the graph to **estimate** the concentration.

Why do you think you used so many concentrations of ONP?

What would you do if your graph was curved, and not linear?

However, to make this experiment more precise, an equation for the line could be determined and you could quantify the product mathematically! Your instructor will show you how to make this graph using Excel and how to generate the equation for the line.

Which of these methods do you feel would be more accurate?

Why?

Remember, you are measuring PRODUCT production, which allows you to estimate ENZYME activity!

NOW, you can use the standard curve to determine the rate of enzymatic reaction and the amount of enzyme present. Obtain two of the tubes labeled "unknown concentration A" and "unknown concentration B." Determine the absorbance of each using the Spec 20. Then, use the standard curve to ***estimate*** the amount of ONP present in the solutions.

Record your data below.

| | Absorbance | Concentration ONP |
|---|---|---|
| Unknown solution A | | |
| Unknown solution B | | |

Now, using Excel, generate an equation for the line and determine the concentration of ONP mathematically.

| | Absorbance | Concentration ONP |
|---|---|---|
| Unknown solution A | | |
| Unknown solution B | | |

Compare the results from your estimate with the results using the linear equation. The values should be similar.

Which of the two solutions contain more product?

## Fun Fact

Without enzymes, our bodies could not function. OMP decarboxylase is an enzyme that helps us make our nucleic acids. It catalyzes a reaction that, without the enzyme, would occur only once every 78 million years! That would be a long time to wait for our DNA and RNA to be made!

## Activity 2—Conducting an Enzyme Assay

Now you will learn to conduct an enzyme assay using an enzyme stock solution that you prepare. The enzyme we will be using is lactase, and is easily obtained from Lactaid tablets, which some people use for lactose intolerance. You will work at your desk as a group of 4 for this experiment.

1. Obtain one Lactaid pill, and crush into a powder with a mortar and pestle.
2. Add the powder to 5 mL 0.1M phosphate buffer (pH7) and vortex to dissolve.
3. Centrifuge 1 mL of this solution at a time in the microcentrifuge tubes provided for 5 minutes. The supernatants (fluid in the top) should be combined into a clean test tube; this will be your enzyme stock solution. The microfuge tubes should look like this following centrifugation. YOUR ENZMYE SHOULD BE KEPT ON ICE FOR THE REMAINDER OF THE EXPERIMENT.

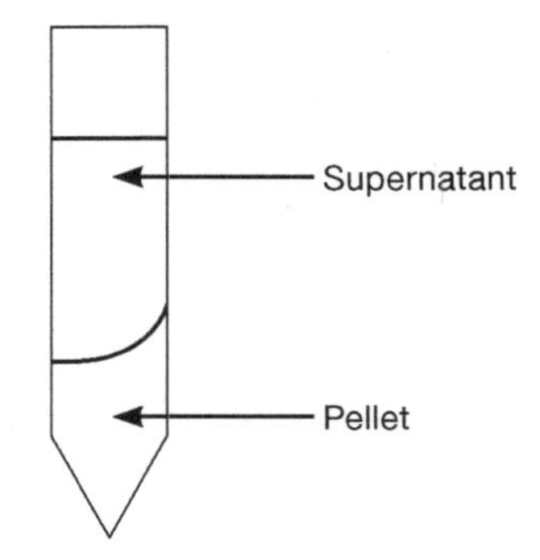


Where is the enzyme? ______

4. Now you will conduct an enzyme assay in a clean test tube. You will need a timepiece available for this assay.
   a. First combine 3.5 mL buffer and 0.5 mL **ONPG—YOUR SUBSTRATE.**
   b. Next obtain 0.5 mL **sodium carbonate = YOUR STOP SOLUTION** in a disposable pipette.
   **Add enzyme when you are ready with the stop solution and stopwatch.**
   c. With the stop solution ready to be added, add 0.5 mL **ENZYME** to the buffer and substrate and **begin timing immediately when you add the enzyme.**
   d. When a pale yellow color appears, **record the time and add the stop solution immediately.**

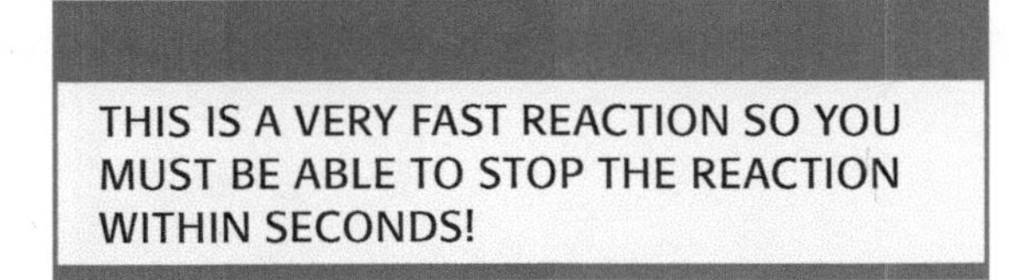


5. After you have stopped the reaction, determine the absorbance using the Spec 20 and record your data in the following table.

| | Start time | End time | Total time (seconds) | Absorbance | Concentration of product (µmoles) | Rate* |
|---|---|---|---|---|---|---|
| Enzyme Assay | | | | | | |

* The rate is the amount of product produced / time.

6. Calculations:
   a. Determine the *concentration of product* in the tube using your standard curve that you generated in Activity 1.
   b. Determine the rate of an enzymatic reaction by dividing the amount of product (ONP in this case) by time. Don't forget your units! Generally speaking, the more product you obtain, the more active your enzyme is.

# Review Questions

**1** Please define the following terms.

a. Enzyme:

b. Substrate:

c. Product:

d. Activation energy:

**2** Why did we use sodium carbonate as a stop solution?

Next week you will examine the effect of various environmental conditions on enzyme activity. You will determine the effect of temperature, pH and enzyme concentration of enzyme activity. For each of these three experiments, please write the independent and the dependent variables:

Does temperature affect enzyme activity?

Dependent variable:

Independent variable:

Does pH affect enzyme activity?

Dependent variable:

Independent variable:

Does enzyme concentration affect enzyme activity?

Dependent variable:

Independent variable:

# Enzymes Part 2

6

## What environmental factors affect enzyme activity?

### Preparation

The first thing you have to do today is to prepare a fresh stock solution of enzyme. Do you remember how to do this? Discuss with your group and let the instructor know when you are ready to proceed.

1. Obtain one Lactaid pill, and crush into a powder with a mortar and pestle.
2. Add the powder to 5 mL 0.1M phosphate buffer (pH7) and vortex to dissolve.
3. Centrifuge 1 mL of this solution at a time in the microcentrifuge tubes provided for 5 minutes. The supernatants (fluid in the top) should be combined into a clean test tube; this will be your enzyme stock solution. The microfuge tubes should look like this following centrifugation. YOUR ENZYME SHOULD BE KEPT ON ICE FOR THE REMAINDER OF THE EXPERIMENT.

You will also need to recall from last week how long you let the enzyme assay run (i.e., how long did it take for color to appear?).

You will use the same time this week.

**Normally, you would generate a new standard curve for the experiment today, but for the sake of time, we will use the standard curve that YOU generated last week.**

## A. The Effect of Temperature on Enzyme Activity

The first experiment will assess the effect of high temperature on enzyme activity. Remember that this enzyme is a protein and discuss with your group what you think the effect of high temperature on this enzyme will be. Do you think that boiling the enzyme will affect the ability of the enzyme to produce product?

Clearly state your hypothesis. ______________________________

______________________________

1. Boil 1 mL of the enzyme stock solution for 15 minutes in a boiling water bath. Be sure to cover the tubes with foil to prevent evaporation. While this is boiling move on to Activity B.
2. Now you will conduct an enzyme assay in a clean test tube. You will need a timepiece available for this assay.
   a. First combine 3.5 mL buffer and 0.5 mL **ONPG – YOUR SUBSTRATE.**
   b. Next obtain 0.5 mL **sodium carbonate = YOUR STOP SOLUTION** in a disposable pipette.
      **Add enzyme when you are ready with the stop solution and stopwatch.**
   c. With the stop solution ready to be added, add 0.5 mL **ENZYME** to the buffer and substrate and **begin timing immediately when you add the enzyme.**
3. Allow the reaction to proceed **for *the same amount of time you used in Activity 2 last week***, then stop with 0.5 mL sodium carbonate.
4. Determine the absorbance using the Spec 20.

5. Fill out the following table.

| Product concentration without boiling (from last week) | Absorbance after boiling | Product concentration after boiling |
|---|---|---|
| | | |

6. Answer the following questions.
   a. Was your hypothesis supported or refuted?

   b. Why or Why not?

   c. What characteristics of the enzyme explain these findings?

## B. The Effect of pH on Enzyme Activity

The next experiment will assess the effect of different pH conditions on enzyme activity. Remember that this enzyme is a protein, and discuss with your group what you think the effect of low pH (pH 2) and a high pH (pH 10) on this enzyme will be. The experiments that you have conducted so far have been at pH 7. Do you think that changing pH will affect the ability of the enzyme to produce product?

Clearly state your hypothesis.

1. For pH 2—Incubate 1 mL of the enzyme solution for 30 minutes in 1 mL of a pH 2 buffer at room temperature.
2. For pH 10—Incubate 1 mL of the enzyme solution for 30 minutes in 1 mL of a pH 10 buffer at room temperature.
3. After 30 minutes, you will conduct two enzyme assays in clean test tubes. You will need a timepiece available for this assay. For each assay:
   a. First combine 3.5 mL buffer and 0.5 mL **ONPG – YOUR SUBSTRATE.**
   b. Next obtain 0.5 mL **sodium carbonate = YOUR STOP SOLUTION** in a disposable pipette.
   **Add enzyme when you are ready with the stop solution and stopwatch.**
   c. With the stop solution ready to be added, add 0.5 mL **ENZYME** to the buffer and substrate and **begin timing immediately when you add the enzyme.**
4. Allow the reaction to proceed **for *the same amount of time you used in Activity 2 last week***, then stop with 0.5 mL sodium carbonate.
5. Determine the absorbance using the Spec 20.

6. Fill out the following table.

| Product concentration at pH 7 (μM) (from last week) | Absorbance at pH 2 | Product concentration at pH 2 (μM) | Absorbance at pH 10 | Product concentration at pH 10 (μM) |
|---|---|---|---|---|
| | | | | |

7. Answer the following questions.
   a. Was your hypothesis supported or refuted?

   b. Why or Why not?

   c. What characteristics of the enzyme explain these findings?

   d. What environment does Lactaid normally work in, and how would that affect your interpretation of these results?

## C. The Effect of Enzyme concentration on Enzyme activity

The next experiment will assess the effect of enzyme concentration on enzyme activity. Do you think that diluting the enzyme will have an effect on the ability of that enzyme to produce product?

Clearly state your hypothesis.

Make a set of 4 **serial dilutions of the enzyme stock solution** and label them as 1:10, 1:100, 1:1,000 and 1:10,000. Serial dilutions are made by diluting the stock solution in a series of tubes, as shown below.

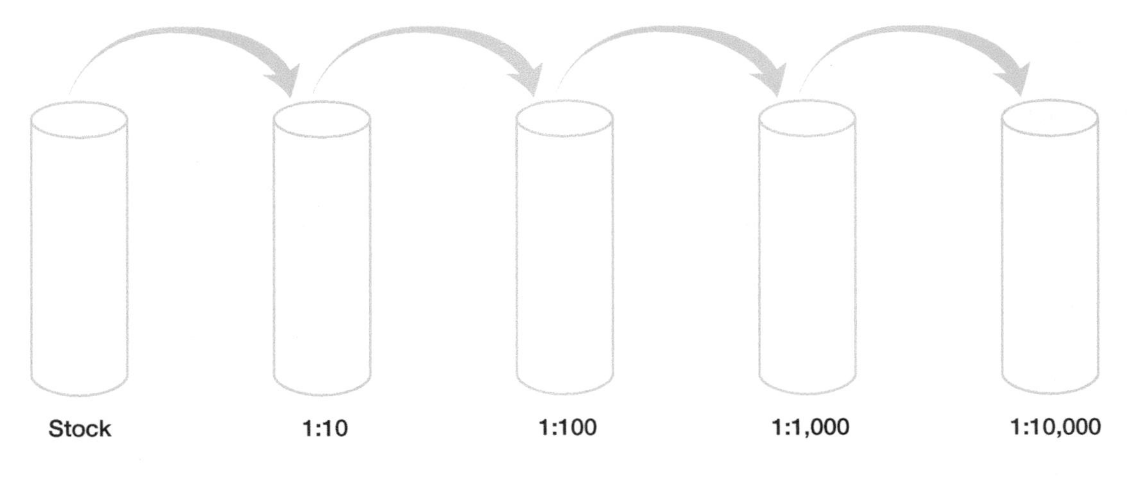

1. Make the serial dilutions by transferring 1 mL of the solution and adding to 9 mL of phosphate buffered saline solution. These dilutions can be made in clean test tubes. Be sure to mix the contents of each tube before transferring the solution to the next test tube.

2. Then, for each dilution, you will conduct an enzyme assay by adding each of the following into a clean test tube:
    a. First combine 3.5 mL buffer and 0.5 mL **ONPG – YOUR SUBSTRATE.**
    b. Next obtain 0.5 mL **sodium carbonate = YOUR STOP SOLUTION** in a disposable pipette.
    **Add enzyme when you are ready with the stop solution and stopwatch.**
    c. With the stop solution ready to be added, add 0.5 mL **ENZYME** to the buffer and substrate and **begin timing immediately when you add the enzyme.**

3. Allow the reaction to proceed **for *the same amount of time you used in Activity 2 last week***, then stop with 0.5 mL sodium carbonate.

4. You should stagger the start of the tubes by at least one minute so that the reactions do not finish at the same time.

5. When all reactions have been stopped, measure the absorbance and determine the concentration of product in each tube using your standard curve. Record your data in the following table.

| | Absorbance | Concentration of product (μmoles) | Rate* |
|---|---|---|---|
| Undiluted | | | |
| 1:10 | | | |
| 1:100 | | | |
| 1:1,000 | | | |
| 1:10,000 | | | |

* The rate is the amount of product produced/time.

6. Answer the following questions.
    a. Was your hypothesis supported or refuted?

    b. Why or Why not?

    c. What characteristics of the enzyme explain these findings?

# Review Questions

You have learned that environmental conditions can affect enzyme activity. Below are some unusual environmental conditions—please discuss where these conditions might be found and what types of organisms (and their enzymes of course) would be found there. How would these conditions affect enzymes from the human body? Remember, every enzyme has an optimal temperature and an optimal pH.

Scenario #1—water with a temp of >100°C

Scenario # 2—liquid with a pH = 2

Scenario #3—a very halophilic (salty) environment

This is a great time to discuss the extremophiles, mostly members of the domain Archaea, and to think about all the life forms different from ourselves! Each of these organisms relies on enzymes too! Here is a place to start: **http://www.youtube.com/watch?v=y3GF3PFNx8Y**

## References:


**http://www.learnnc.org/lp/pages/3398**

**http://bcrc.bio.umass.edu/intro/manual/index.php?title=enzyme_activity&oldid=2535**

# Mitosis and Meiosis

7

## *Before coming to lab*

- ☐ Read Meiosis Lab pages 70–76 and Meiosis Lab pages 77–78
- ☐ Answer the questions on page 70 ***before coming to the Mitosis and Meiosis Lab.***

# Splitting Up
## Understanding Cell Division and Meiosis

*Sit down before fact as a little child, be prepared to give up every conceived notion, follow humbly wherever and whatever abysses nature leads, or you will learn nothing.*

—Thomas Henry Huxley (1825–1895)

## OBJECTIVES

*At the completion of this chapter, the student will be able to:*

1. Discuss the functions of cell division.
2. Describe the cell cycle and how it is controlled.
3. Distinguish and describe the three stages of interphase and the stages of mitosis.
4. Draw and label the stages of interphase and mitosis.
5. Compare and contrast mitosis and meiosis.
6. Explain the relevance of meiosis to sexual reproduction.
7. Describe how chromosomes are reduced from diploid number (2*n*) to haploid number (*n*) in meiosis.
8. Describe the process of tetrad formation, synapsis, and crossing over.
9. Compare and contrast spermatogenesis and oogenesis.

*Answer the following questions prior to coming to lab.*

**1** What are the major types of cancer and their incidence?

______________________________

______________________________

**2** What are some new treatments for cancer?

______________________________

______________________________

**3** What is trisomy 21 and some of its characteristics?

______________________________

______________________________

**4** What is the significance of crossing over?

______________________________

______________________________

**5** Is there any relationship between mother's age and the incidence of trisomy 21?

______________________________

______________________________

Many times the terms **mitosis** and **meiosis** are confused because they sound so much alike. Ultimately both processes are essential to the proliferation of life on Earth and share many fundamental steps. However, each process is distinctly different. Mitosis is a process that occurs during cell division, which results in a parent cell giving rise to two identical **daughter cells**. Meiosis, on the other hand, results in the formation of **gametes,** or sex cells, through a series of reduction divisions involving the chromosomes. Keep in mind that life cycles can be complex and that in some lower plants and fungi meiosis may produce asexual spores.

# Cell Division

Many questions arise when discussing cell division. How can a one-celled human zygote (fertilized egg) grow to eventually become an adult consisting of approximately 125 trillion cells? How can a bacterial infection spread so quickly? How can a cancerous tumor grow so quickly? How do those pesky weeds seem to pop up overnight in your yard? These are a few of the many questions that can be explained with an understanding of the amazing process known as cell division.

Living organisms as well as the cells that compose tissues are capable of reproduction and growth. **Cell division** is the mechanism by which new cells are produced, whether for growth, repair, replacement, or forming a new organism. The concept of cell division was described in 1855 by the German scientist Rudolph Virchow (1821–1902) when he declared that "cells come from pre-existing cells."

In 1875, German botanist Eduard Strasberger (1844–1912) first described cell division in plants, but the processes of cell division were not described in detail until 1876. In that year, German zoologist Walther Flemming (1843–1905), when describing the development of salamander eggs, coined the terms **chromatin** (from the Greek word for colored or dyed) and **mitosis** (from the Greek word for thread) and also established the framework for understanding the stages of cell division. Today, mitosis is recognized as a major component of the overall process of cell division.

Cell division is the biological process by which cellular and nuclear materials of **somatic cells** (nonsex cells) are divided between two daughter cells formed from an original parent cell. The resulting daughter cells are structurally and functionally similar to each other and to the parent cell. In prokaryotic organisms (Bacteria and Archaea) the distribution of exact replicas of genetic material is comparatively simple. In eukaryotic organisms, such as animals, the process of cell division is much more complex. The complexity is the result of the presence of a larger cell, a nucleus, and more DNA (larger genome) on individual linear chromosomes. Thus, any study of cell division in eukaryotes will include a discussion of the **cell cycle** and its two components, **interphase** and mitosis.

Dividing cells pass through a regular sequence of cell growth and division known as the cell cycle (Fig. 7.1). **Checkpoints** in the cell cycle ensure that cell division is occurring properly. The first checkpoint, which occurs at the end of $G_1$, monitors the size of the cell and whether the DNA has been damaged. If a problem is detected, either repair or **apoptosis** (programmed cell death) occurs. At the second checkpoint, which occurs at the end of $G_2$, the cell will proceed to mitosis only if the DNA is undamaged and if DNA replication has occurred without error. The mitotic spindle checkpoint during metaphase, the final checkpoint monitors the spindle assembly and controls the onset of anaphase. Cell types, hormones, and growth factors as well as external conditions influence the time required to complete the cell cycle. The cell cycle is divided into two major stages: interphase (when the cell is not actively dividing) and mitosis (when the cell is actively dividing). At any given time, a cell exists in one of the stages of the cell cycle. The majority of cells in an organism are in interphase.

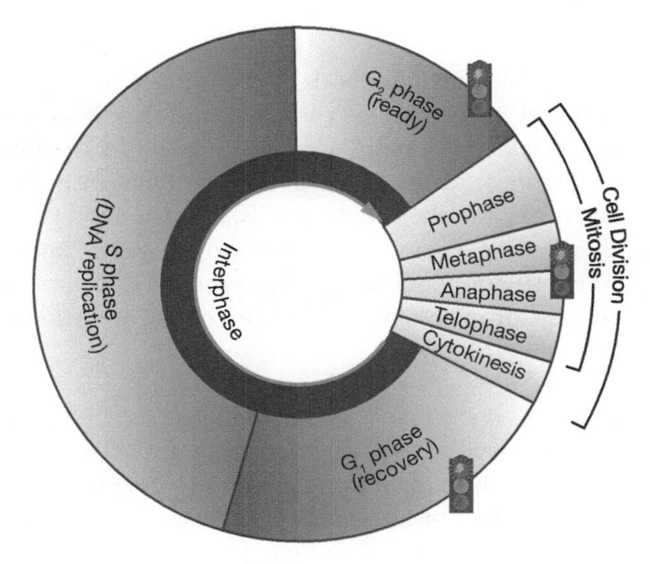


FIGURE **7.1** Cell cycle.

## Interphase

Interphase typically accounts for 90% of the time that elapses during each cell cycle. Classically called the *resting stage*, interphase is actually a busy and extremely important part of the cell cycle. During interphase and before a cell begins the process of mitosis, it must undergo DNA replication, synthesize important proteins, produce enough organelles to supply both daughter cells, and assemble the structures used during cell division.

Interphase is divided into three phases: gap 1 ($G_1$), synthesis (S), and gap 2 ($G_2$). In recent years, a period known as $G_0$ has been described. $G_0$, or the quiescent phase, occurs after $G_1$ and represents the time during which a cell is metabolically active but not proliferative. Actively dividing cells and cancer cells either skip $G_0$ or pass through the stage quickly. Some cells, such as nerve cells, may never exit $G_0$ once formed.

The $G_1$ phase, occurring after mitosis, serves as the cell's primary growth phase. During this time, the cell recovers from the previous division, increases in size, and synthesizes proteins, lipids, and carbohydrates. Also, the number of organelles and inclusions increases in number. In cells that contain **centrioles**, the two centrioles begin to form during $G_1$ (note that the cells of flowering plants, fungi, and roundworms do not contain centrioles). The $G_1$ phase occupies the major portion of the life span of a typical cell. Slow-growing cells, such as some liver cells, can remain in $G_1$ for more than a year. Fast-growing cells, such as epithelial cells and those of bone marrow, remain in $G_1$ for 16–24 hours.

In the S phase of interphase, which follows the $G_1$ phase, each chromosome replicates to produce two daughter copies, or **sister chromatids** (Fig. 7.2). The two copies remain attached at a point of constriction called the **centromere**. During this time, these structures are not visible under the light microscope. Among the numerous proteins manufactured during this phase are histones and other proteins that coordinate the various events taking place within the nucleus and cytoplasm. Duplication of the centrioles is completed, and they organize to migrate to the opposite poles of the cell. In addition, the microtubules that will become part of the spindle apparatus are synthesized.

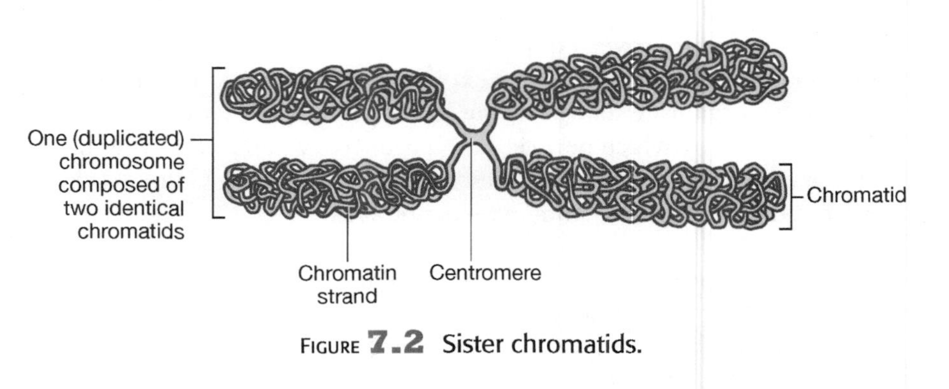


FIGURE 7.2 Sister chromatids.

The $G_2$ phase of interphase, which occurs after the S phase, involves further replication of membranes, microtubules, mitochondria, and other organelles. In addition, the newly replicated sister chromatids diffusely distributed throughout the nucleus begin to coil and become more compact. The start of chromosome condensation at the completion of $G_2$ signals the beginning of mitosis. By the end of $G_2$, the volume of the cell has nearly doubled.

## Mitosis and Cytokinesis

Cell division includes both the division of the genetic material, mitosis, or **karyokinesis**, and the division of the cytoplasm, **cytokinesis**. The overall goal of karyokinesis is to distribute identical sets of chromosomes from the parent cell to each of the daughter cells. Keep in mind that cell division is a dynamic and continuous process, but for ease of explanation and study it has been divided into several stages—prophase, metaphase, anaphase, and telophase.

### Prophase

**Prophase** is the first, longest, and most active stage of mitosis, characteristically accounting for 70% of the time a cell spends in the mitotic process. During this stage, the chromosomes progressively become more visible as they appear to shorten and thicken. Upon close examination in late prophase, the sister chromatids and their centromeres can be seen easily. Prophase also is characterized by migration of the centriole pairs toward opposite poles and disintegration of the nuclear envelope and nucleolus.

> *Did you know . . .*
>
> **The Cell Cycle and Cancer**
>
> Sitting in a crowded stadium can be a humbling experience when you begin contemplating cancer statistics. It is estimated that one in three Americans will develop cancer in their lifetime—just look around. Actually, cancer is defined as a group of diseases characterized by the uncontrolled growth and spread of abnormal cells. In other words, cancer involves cells that have lost control of the cell cycle. It is estimated that there are well over 100 different types of cancer. Unfortunately, many types of cancer are lethal and one in four deaths in the United States is due to cancer. One focus of cancer research is to help cells regain control of the cell cycle.

Short tubules known as **asters** appear and begin to radiate from the centrioles. Asters are believed to stiffen the point of microtubular attachment during the retraction of the spindle. Polar microtubules between the centrioles begin to form the **spindle fibers** that eventually will expand from one pole to another, meeting at the equatorial plane.

The term **prometaphase** has become increasingly popular to describe events of late prophase. It usually is distinguished by the attachment of sister chromatids to the spindle fibers via a proteinaceous hook, or **kinetochore**.

### Metaphase

Metaphase is the brief second stage of mitosis, when the centromere joining each pair of sister chromatids is attached to the spindle. Eventually, the pairs of sister chromatids appear to align midway between the centrioles along what is called the **equatorial plane**, or **metaphase plate**. Early events of metaphase also can be considered in late prometaphase. Cells can be halted during metaphase for cytogenetic studies such as karyotyping and cancer research.

### Anaphase

**Anaphase** is the third and most intriguing stage of mitosis. Although brief, this stage is characterized by the separation of the sister chromatids from their centromere and the movement of the chromatids as they are pulled back along the spindle to the opposite poles. Errors during anaphase could result in an unequal distribution of genetic material with devastating consequences.

### Telophase

In the final stage of mitosis, **telophase**, the cell resembles a dumbbell with a set of chromosomes at each end. During telophase, the spindle is disassembled, the nuclear envelope and nucleolus re-form, and cellular inclusions and organelles organize. Cytokinesis, the division of cytoplasm upon completion of nuclear division, begins during this stage. During cytokinesis in animal cells, a cleavage furrow develops, and the cell appears to be pinched in two. Eventually the cleavage furrow is completed, and two daughter cells are formed. Upon completion of telophase and cytokinesis, the daughter cells enter the $G_1$ phase, and the cell cycle repeats. In plant cells, instead of undergoing cytokinesis, a **cell plate** is formed, dividing the mother cell into two daughter cells.

## Activity 1—The Cell Cycle

### Procedure 1
Animal Cell Division

**Materials**
- ❑ Compound light microscope
- ❑ Prepared slide of whitefish blastula
- ❑ Colored pencils

In this procedure, you will observe the cell cycle in prepared slides of a whitefish blastula (Fig. 7.3). The **blastula** occurs in the early stage in the embryonic development of an animal and appears as a ball of cells, each cell in one of the stages of interphase or mitosis. The blastula can be thought of as a basketball with each pebble on the basketball representing a cell.

1. Procure a prepared slide of a whitefish blastula. Obtain a compound light microscope, and follow the safety and handling procedures for a light microscope.
2. Observe the whitefish blastula first on low power. In this field you will be able to see many slices of depth of the blastula. Focusing on low power, you should see individual cells. Switch to high power, and sketch and describe each of the stages of interphase and mitosis in the space provided below and on the following page. Draw each stage of cell division, and describe what is happening within the cell in detail.

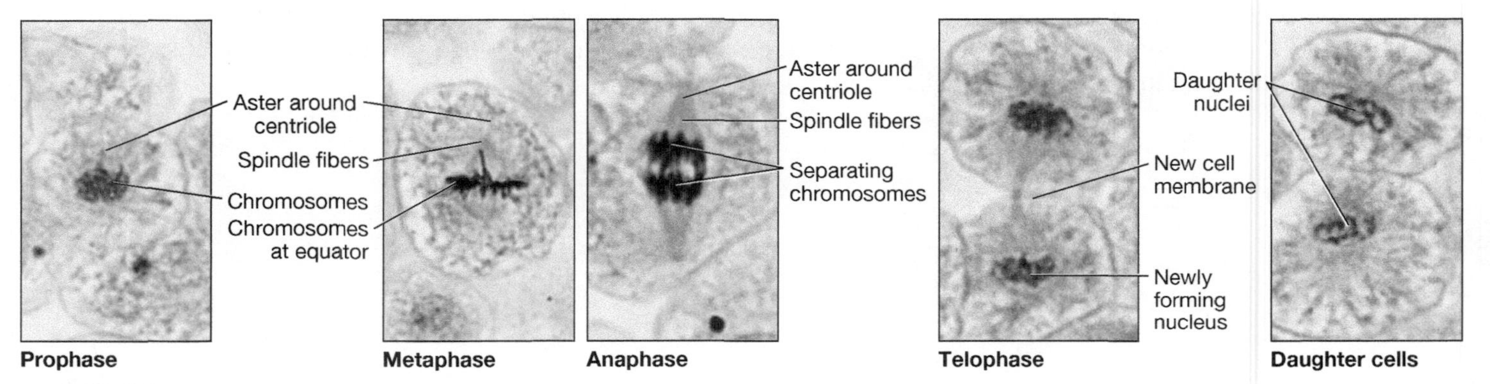


FIGURE 7.3 Stages of animal cell (whitefish blastula; all 500×) mitosis followed by cytokinesis.

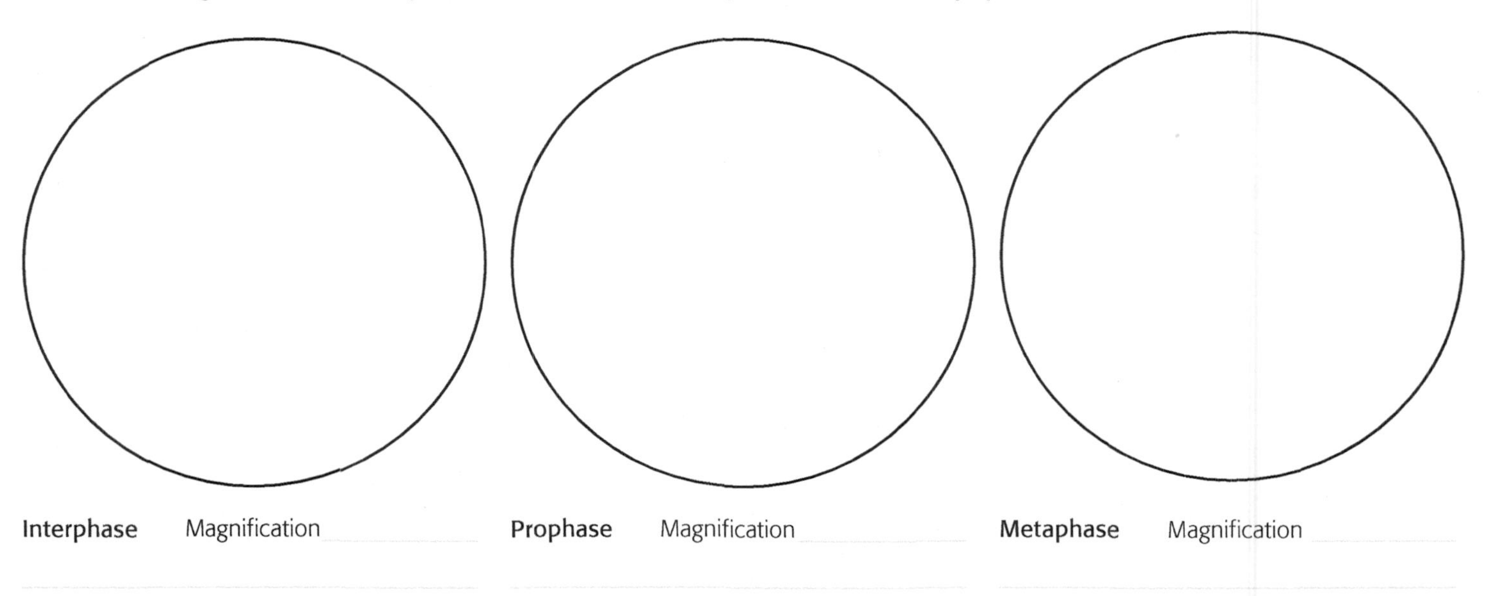

Interphase Magnification ____________

Prophase Magnification ____________

Metaphase Magnification ____________

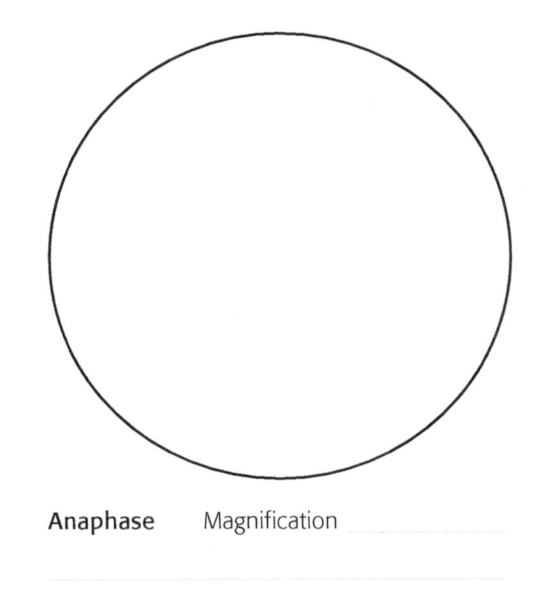

**Anaphase** Magnification

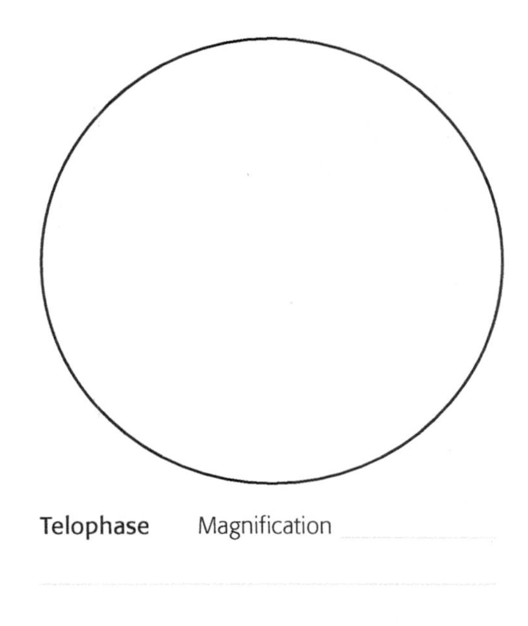

**Telophase** Magnification

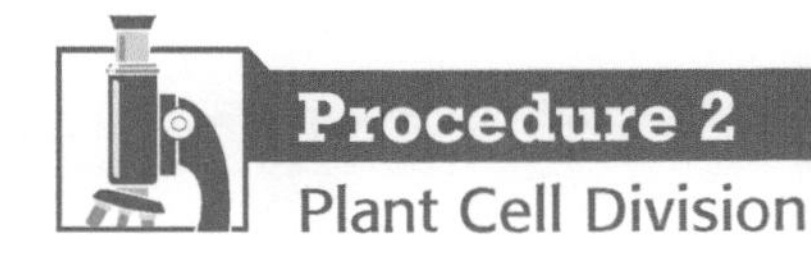

## Procedure 2

### Plant Cell Division

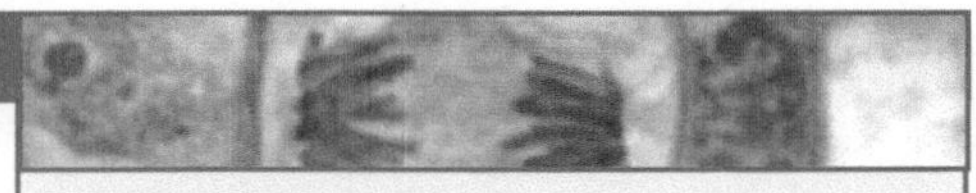

**Materials**

- ❑ Compound light microscope
- ❑ Prepared slide of onion root tip mitosis
- ❑ Colored pencils

In this procedure, you will observe the cell cycle in prepared slides of green onion (scallions) root tip mitosis. In plants, cell growth occurs in the **meristematic** regions, primarily located at the tips of stems and roots, so the root tips often are used to study the cell cycle (Fig. 7.4).

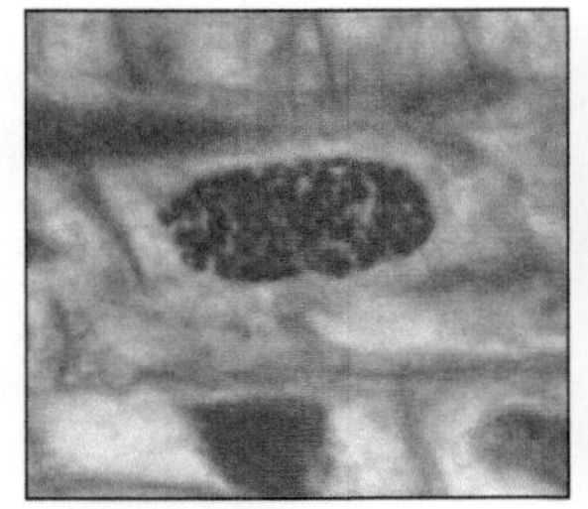

**Early prophase**
Chromatin begins to condense to form chromosomes.

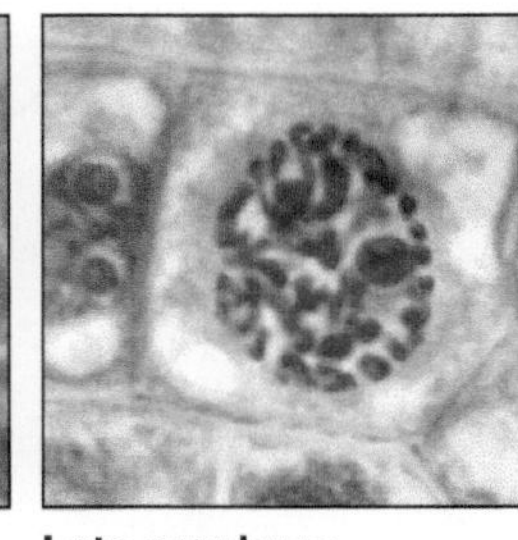

**Late prophase**
Nuclear envelope is intact, and chromatin condenses into chromosomes.

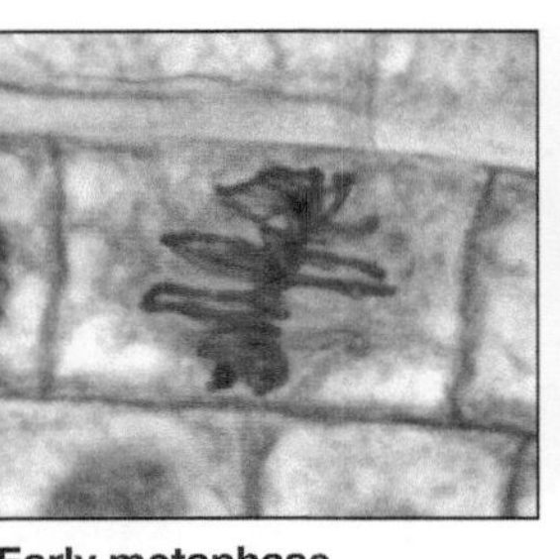

**Early metaphase**
Duplicated chromosomes are each made up of two chromatids, at equatorial plane.

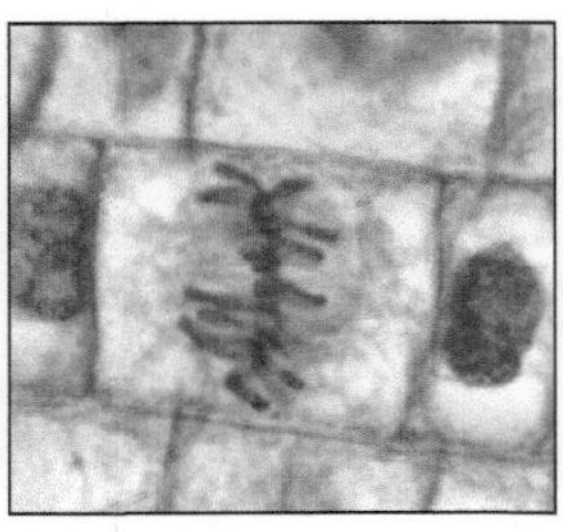

**Late metaphase**
Duplicated chromosomes are each made up of two chromatids, at equatorial plane.

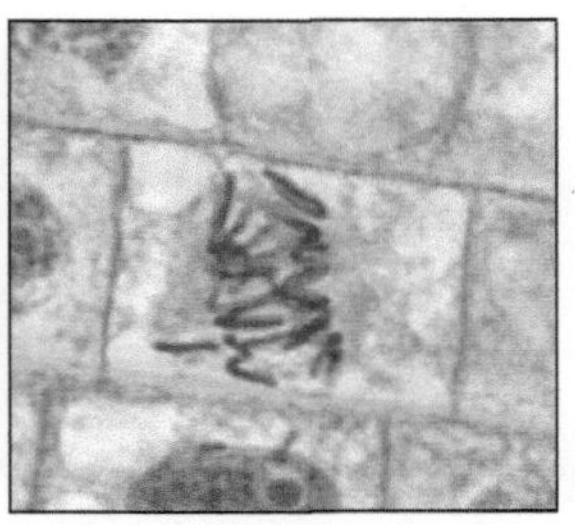

**Early anaphase**
Sister chromatids are beginning to separate into daughter chromosomes.

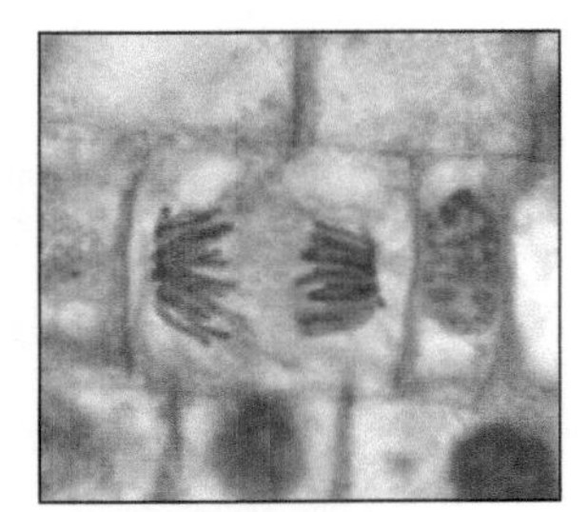

**Late anaphase**
Daughter chromosomes are nearing poles.

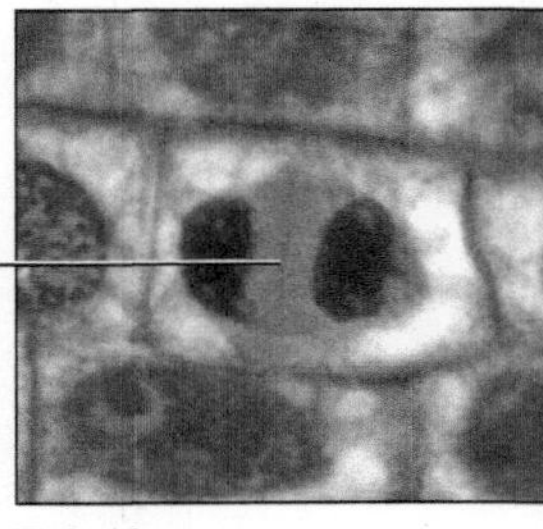


**Telophase**
Daughter chromosomes are at poles, and cell plate is forming.

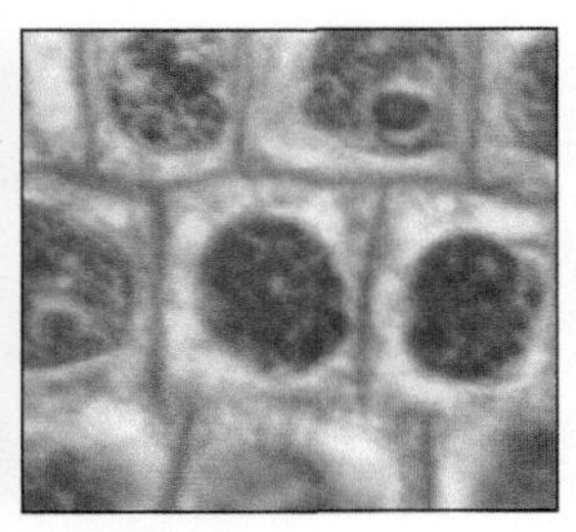

**Interphase**
Two daughter cells result from cytokinesis.

FIGURE **7.4** Stages of mitosis in Hyacinth, *Hyacinthus*, root tip (all 430×).

1. Procure a prepared slide of onion root tip mitosis. Obtain a compound light microscope, and follow the safety and handling procedures for a light microscope.
2. Observe the onion root tip mitosis first on low power. In this field, you will be able to see many stages of interphase and mitosis. Focusing on low power, you should see individual cells. Switch to high power, and sketch and describe each of the stages of interphase and mitosis in the space provided below.

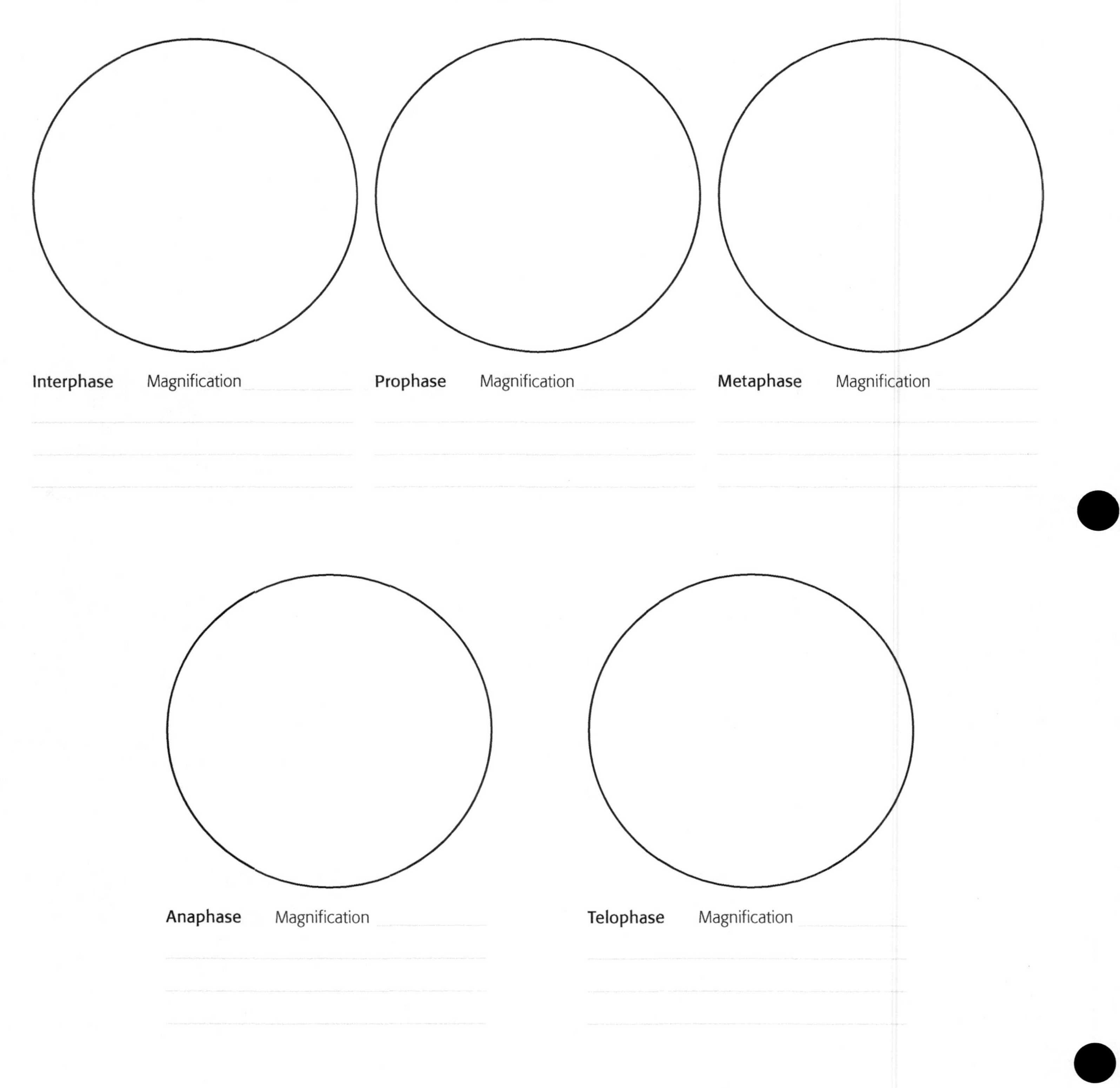

# Meiosis

In addition to the evolution of aerobic respiration, one of the most significant events in the history of life was the evolution of sexual reproduction. Sexual reproduction is responsible for the great diversity of life on Earth. Offspring are no longer "chips off the old cell" but, rather, unique individuals with characteristics inherited from past generations along with new, hopefully beneficial, traits.

The underlying mechanism behind sexual reproduction is **meiosis.** Unlike mitosis, which involves a division of the somatic, or body, cells, meiosis involves reduction division resulting in the formation of sex cells, or **gametes.** This is achieved in two distinct nuclear divisions, **meiosis I** and **meiosis II.** In addition, the genetic recombination that occurs in meiosis generates genetic variation.

The size or complexity of an organism is not determined by the number of chromosomes. With a few exceptions to the normal condition, species possess a genetically determined number of chromosomes. Eukaryotic chromosomes occur in pairs called homologs, or homologous pairs. Homologous chromosomes contain the same genes in the same order along the chromosome; the DNA sequences are not necessarily identical. An individual receives one of each pair from each parent. The **diploid number** ($2n$) represents the total number of chromosomes in the nucleus of a cell. In humans, the diploid number is 46. Example diploid numbers from the living world include: Adder's tongue fern (1,260), goldfish (94), dog (78), horse (64), chimp (48), bottle-nosed dolphin (44), house mouse (40), cat (38), corn (20), and fruit fly (8). The number of pairs of chromosomes is the **haploid number** ($n$), one-half of the diploid number. In humans, the haploid number is 23, and in a fruit fly it is 4. Meiosis involves the reduction of the diploid state ($2n$) to the haploid state ($n$). These haploid cells serve as gametes, or sex cells, such as the sperm and ovum (egg cell). Meiosis also ensures the stability in the number of chromosomes passed from one generation to the next.

Sexual reproduction is based upon **fertilization,** the union of haploid male and female gametes to form a diploid **zygote,** or fertilized egg. The zygote possesses homologous chromosomes from each parent. The process of meiosis varies slightly in the kingdoms Plantae, Fungi, and Animalia.

**Preformation!**

The idea of "where do we come from" confounded early scientists. Perhaps Pythagoras (circa 530 BC) was the first to state that the father contributed his essence to the offspring and the mother contributed the material substrate. The individuals that supported this idea were known as spermists and included a number of well-known scientists through the centuries (Galen, Leeuwenhoek, and von Baer). In 1664, Nicolaas Hartsoeker actually described a little human (homunculus) curled up inside the sperm ready to enter the egg (see figure to the right). On the other hand, ovists such as William Harvey conjectured that the egg contained the homunculus. The ovists contended that women carried eggs containing boy and girl children, and that the gender of the offspring was determined before conception. The battle between the spermists and the ovists and whether we were preformed raged for centuries. So you think that meiosis is complicated!

Meiosis consists of two distinct cell divisions designated meiosis I and meiosis II. Interphase is completed prior to meiosis I.

## Meiosis I

The first stage of meiosis I is prophase I. In this stage, homologous chromosomes pair up in a process called **synapsis** and form **bivalents** or **tetrads.** A synaptonemal complex is formed between homologous chromosomes to ensure proper pairing. The physical exchange of genetic material between homologous pairs known as **crossing over** results in the formation of new combinations of genetic material, making that sex cell unique (Fig. 7.5). The arms of the sister chromatids are held together by chiasmata.

In metaphase I the bivalents line up independently along the metaphase plate (Fig. 7.6). This independent alignment of sister chromatids allows for the genetic diversity seen in families. In contrast to metaphase during mitosis, the sister chromatids are lined up in a double row rather than a single row.

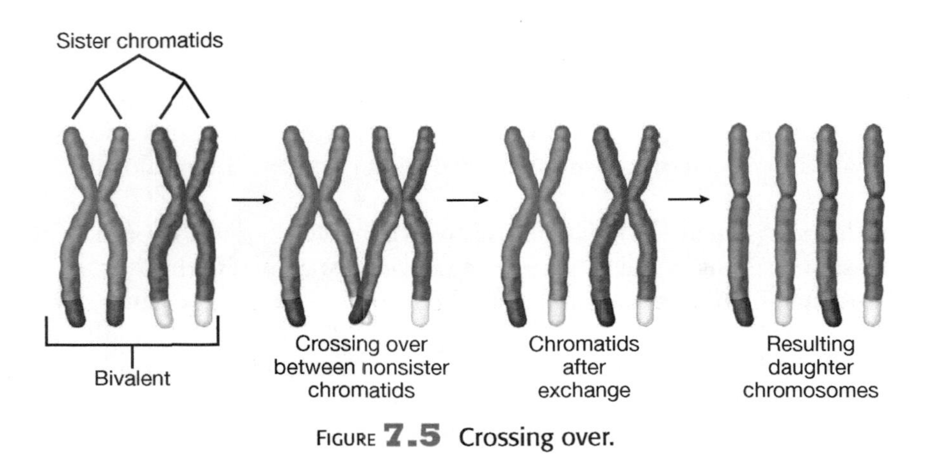


FIGURE 7.5 Crossing over.

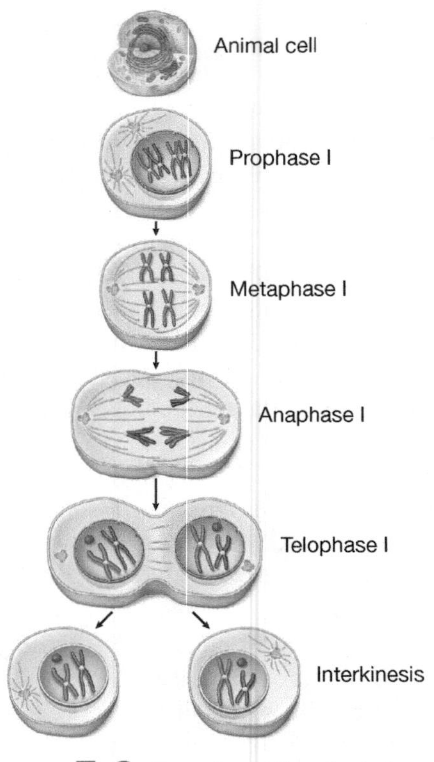


FIGURE 7.6 Overview of meiosis I in animal cells.

Anaphase I follows metaphase I. In this stage, homologous chromosomes separate and migrate toward opposite poles.

Telophase I, the final stage of meiosis I, is characterized by the sister chromatids reaching the opposite poles and eventually forming two new cells. Although the two cells produced at the end of meiosis I are considered haploid by convention, they possess pairs of homologous chromosomes.

In some species, **interkinesis** follows telophase I. It is similar to interphase in the cell cycle, but DNA replication does not occur.

## Meiosis II

Meiosis II is similar to mitosis and results in division of the sister chromatids from meiosis I. Prophase II is characterized by attachment of the sister chromatids to the spindle. During metaphase II, the sister chromatids line up along the metaphase plate. In anaphase II, the sister chromatids are separated and migrate toward the centrioles. Telophase II results in the formation of four unique daughter cells (Fig. 7.7).

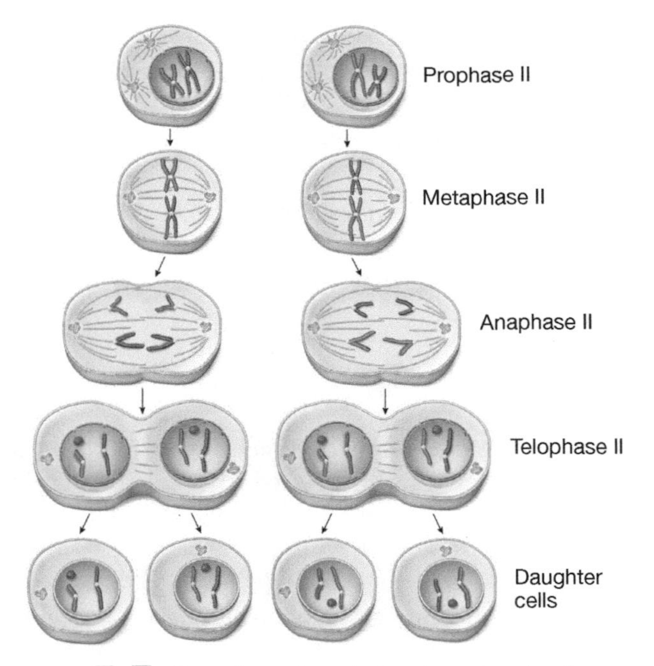


FIGURE 7.7 Overview of meiosis II in animal cells.

## REVIEW EXERCISE—PLEASE COMPLETE THE FOLLOWING TABLE

| | Mitosis | Meiosis |
|---|---|---|
| How many chromosomal duplications? | | |
| How many cell divisions? | | |
| What are the products? | | |
| Are the products haploid or diploid? | | |
| Are the products identical to the parent? | | |
| Where does this process occur in the body? | | |
| What are the functions in the body? | | |

## Activity 2—Comparison of Cancer and Normal Cells

One of the most interesting things about studying cell division is its relationship to cancer. The term cancer represents a group of diseases characterized by uncontrolled cell growth. Cancer cells often look different than the normal cells, and pathologists use these morphological differences to detect cancer cells. In this experiment, you will view them under the microscope to visualize the differences between cancer and non-cancer cells.

### Introduction

Early studies on transformed malignant cells established that cancer-forming cells exhibit unique and distinguishing features in comparison to their normal cell counterparts. First, the cells were capable of sustained growth in culture that were virtually immortal. Second, the cells overgrew in culture forming a mass due to piling of cells. Third, solid tumor cells growing in culture were less adherent to each other and the tissue culture substratum. Fourth, the cells' morphologic appearance ranged from resembling the normal cell from which they arose to an embryonic form. Finally, most show an increase in number of nuclei, and distinct karyotype including translocation, deletions, and chromosomal inversion.

Much of the current work is focused on how genetic alterations promote the selections of cancer cells with aggressive behavior. Two classes of genetic alterations occur in cancer including the gain of oncogenes and the loss of tumor suppressors. Proto-oncogenes code for normal proteins in a cell that are involved in accelerating growth. Oncogenes arise in cancer cells when proto-oncogenes mutate or amplify leading to enhanced activity promoting cell growth. An analogy of how oncogenes act would be that of an automobile with the gas pedal constantly pushed down to increase the speed of a car. Tumor suppressors, otherwise known as anti-oncogenes, are equally important in the development of cancer. The normal function of tumor suppressor proteins is to restrict cellular proliferation providing a protective function. When these genes undergo mutagenesis or inactivation, cellular proliferation continues without regulation, leading to cancer growth. Again, the analogy of how tumor suppressors act would be that the automobile's brake pedal slows or regulates the rate of a cell's division. In cancer, the oncogene takes over cell division and the altered or inactive suppressor no longer has control over the process. In almost all cases the functional consequence of these genetic changes is that cancer cells exhibit increased growth, altered cell surface properties, loss of normal cellular morphology and/or resistance to anti-cancer drugs, and increases in number of nuclei and aberrant karyotype.

Although different cancer types, as well as cancer cells from a single tumor mass, may show a different genetic makeup and properties, the distinguishing feature of cancer cells is their loss of growth control. Cancer cells show an abnormally high growth rate. This property of cancer cells is the basis of how anti-cancer drugs selectively kill rapidly growing cancer cells but do not harm slow growing normal cells with in a patient's body. Cell growth involves an increase in cell mass and size which then triggers a cell to divide. Thus, cell growth and cell division are linked. Cell division involves two recognizable coordinated events; the duplication of cell DNA and physical division of the cells into two daughter cells. Because cancer cells replicate so fast, they often show a multinucleated phenotype and contain numerous nuclei, in addition, cancer cells often show an abnormal variability in the size and shape of their nuclei. Pathologists often use the abnormal nuclei observed in cancer cells to diagnose cancer.

What contributes to the uncontrolled growth of cancer cells? Normal cells show a property called contact inhibition, whereby cell growth ceases under conditions of decreased nutrients, injury or cell crowding. Cancer cells, however, have lost contact inhibition and continue to grow when normally cells cease to grow. This increased growth leads to cancer cells colonizing and destroying normal tissue. It is thought that defects in the cell membrane of cancer cells interfere with the ability of these cells to "sense" the cellular environment.

In addition to alterations in cell growth and loss of contact inhibition, cancer cells show altered cell shape. Thus, cancer cells look different from their normal counterparts. Normal cells grow as ordered patterns as the cell density increases. In contrast, cancer cells form chaotic masses. One consequence of their altered morphology of cancer cells is that these altered cell shapes contribute to increased cell movement seen in cancer cells. A direct consequence of this defect leads to malignant cancer cells migrating and spreading throughout the patient's body.

The experimental objective of this lab is to observe morphological changes in cancer cells as compared to normal cells.

## Observations and Questions

Cancer cells are said to have the following characteristics. Describe in your own words what each characteristic means.

1. Cancer cells do not exhibit *density inhibition.*

2. Cancer cells do not exhibit *anchorage dependency.*

3. Cancer cells exhibit *uncontrolled cell division.*

4. Cancer cells are a disease of the *cell cycle checkpoints.*

Observe the prepared slides of normal lung tissue, and the adenocarcinoma lung tissue. Draw the tissue the best that you can in the spaces below:

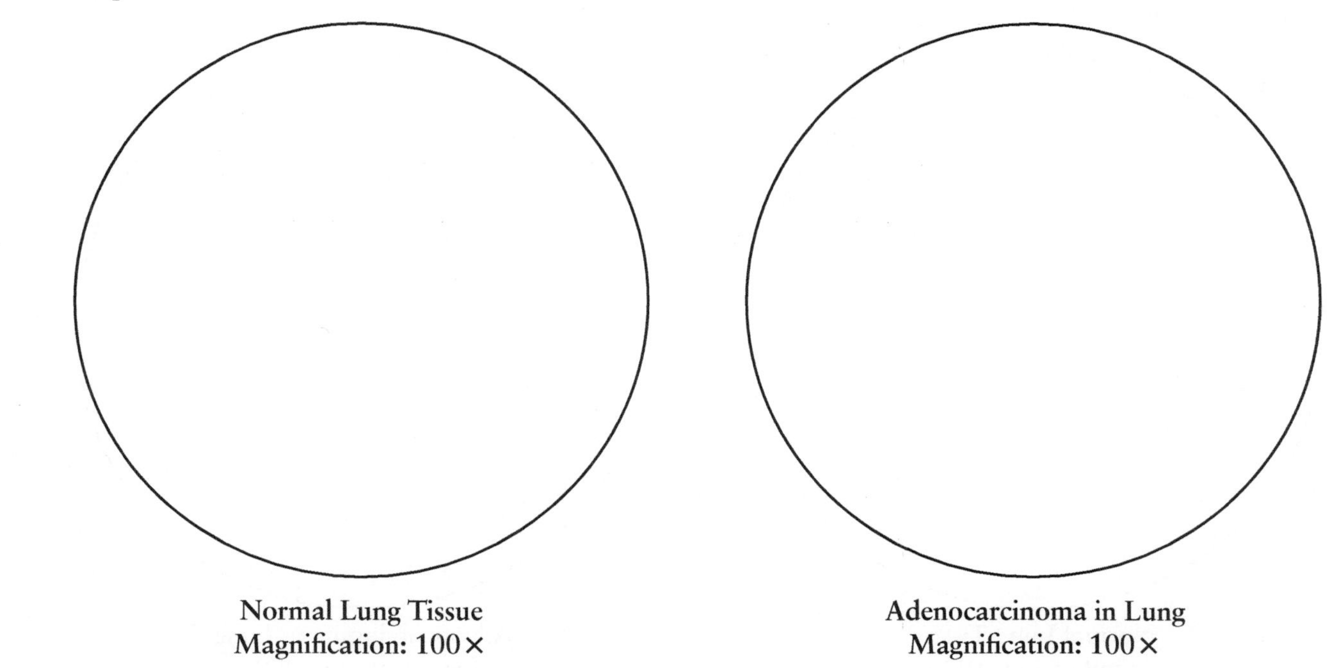

**Normal Lung Tissue**
**Magnification: 100×**

**Adenocarcinoma in Lung**
**Magnification: 100×**

5. Do a quick internet search and describe the function of healthy lung tissue.

6. How does the structure of the cells in the prepared slide of the normal lung tissue match the function described above?

7. Define *carcinoma*.

8. Based on what you now know about healthy lung structure, how does the appearance of the adenocarcinoma lung tissue lead you to believe that it might impede normal lung functioning?

9. The introduction of this lab describes the characteristics of cancer cells. Based on the oil immersion demonstration, *identify* and *draw* the characteristics of normal cells and cancer cells that you can observe directly from the comparison of these two slides.

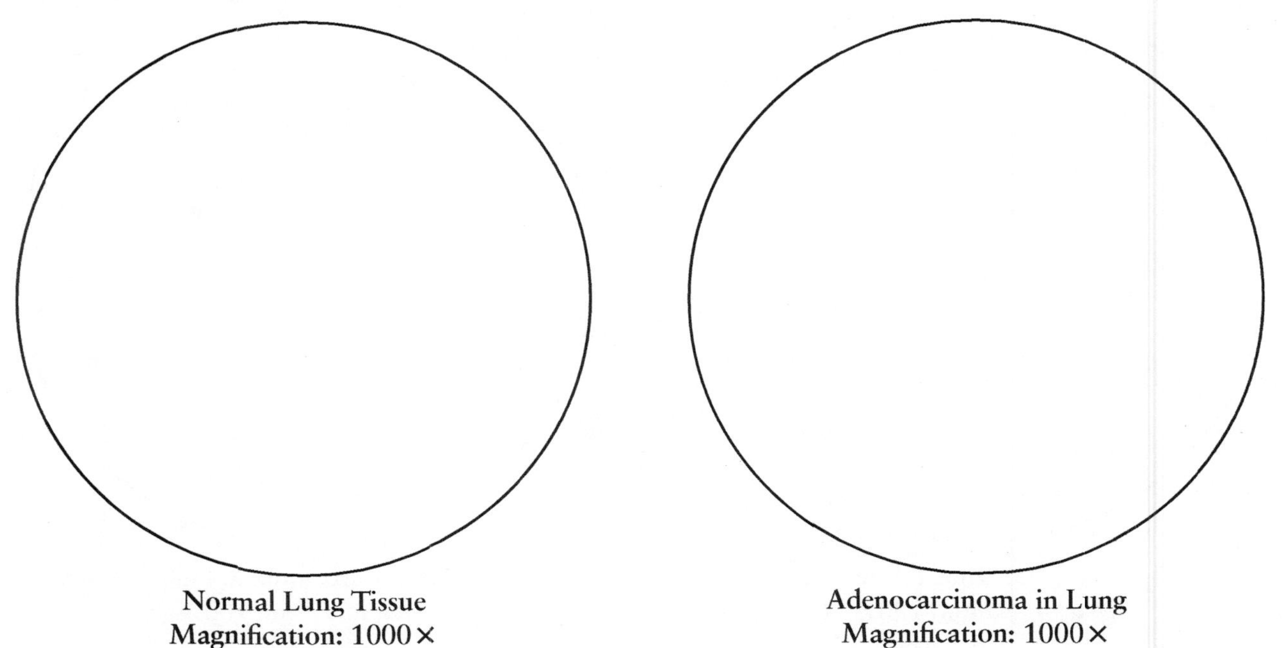

**Normal Lung Tissue**
**Magnification: 1000×**

**Adenocarcinoma in Lung**
**Magnification: 1000×**

List the differences you see between these cells:

# DNA Technology

8

***Before coming to lab***

- ☐ Read lab
- ☐ Answer pre lab questions
- ☐ Define the genes listed on page 91

## PRE LAB QUESTIONS

1. What chemicals make up DNA?

2. Explain the base pairing rules that govern DNA structure.

3. What is the charge of a DNA molecule?

4. What are restriction enzymes?

5. What is microarray technology?

6. What is cDNA?

7. What is an oncogene?

**Before beginning the DNA Technology Lab, examine the results of the Fast Plant Experiment you started last week.**

## Purpose of the lab

The purpose of today's lab is to introduce the student to several powerful molecular biology techniques that are currently being used to answer important scientific questions. The techniques that we will study today are DNA fingerprinting, which involves restriction digests and gel electrophoresis, and microarray analysis. These are just two of the myriad molecular techniques being used today. These experiments should prompt a discussion of the limits of our technologies, and the ethics surrounding these technologies.

## Introduction—DNA Fingerprinting

If you remember, we isolated your own DNA just a few weeks ago. The slimy substance you isolated is actually made up of just a few chemicals, namely four nucleotide bases (adenine, guanine, cytosine and thymine), phosphate and a sugar called deoxyribonucleic acid. DNA exists as a double helix, in which the nucleotide bases are held together by hydrogen bonds. Adenine always pairs with thymine and cytosine always pairs with guanine. The backbone of the molecule is made of an alternating chain of phosphates and sugar molecules. Here are three different representations of a DNA molecule: **Label the bases, sugars and phosphates on the first panel.**

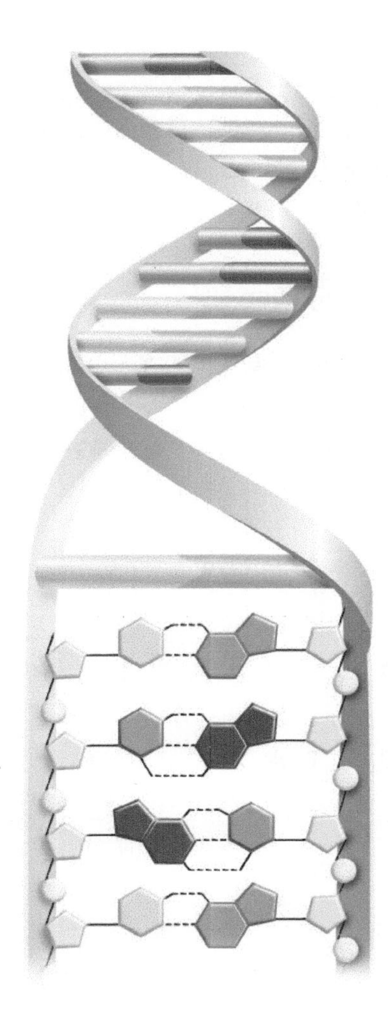

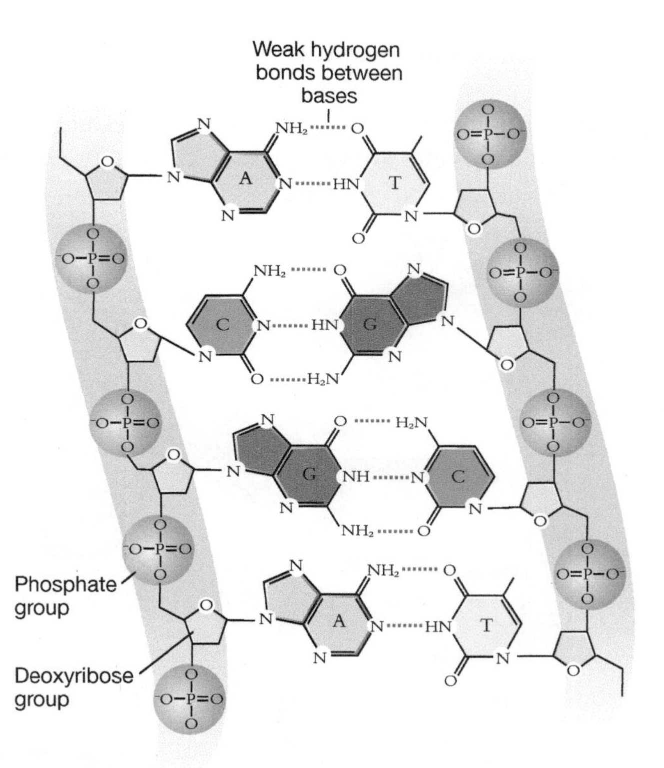

Each of our cells contains an enormous amount of DNA, and a great deal of information can be learned about an organism from its DNA. The reason we are all unique, and the reason for the vast diversity among living organisms, is due to the *order of the base pairs* in our DNA. Today we are going to look at the order of base pairs in the DNA of a virus to determine the cause of a deadly foodborne outbreak. This will be simulated—no real pathogenic viruses are being used!

Many viruses use DNA as their genetic material, just like we do. A typical virus might look like this:

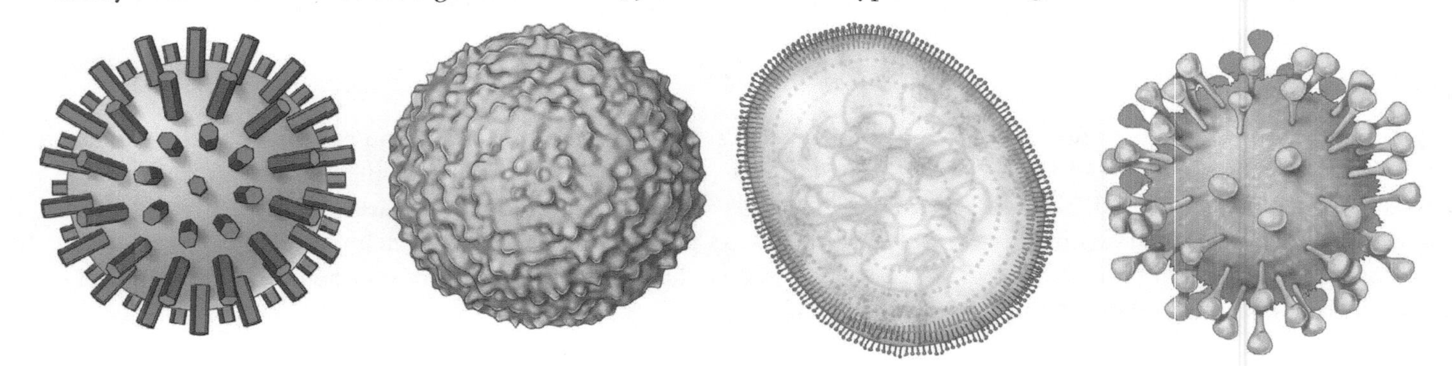

Their DNA is contained within a protein capsid. Just like in humans, the sequence of their DNA determines what they can and cannot do. Scientists today use many techniques to determine the order of the bases in DNA to learn about an organism. Sometimes, the DNA from an organism will be isolated and then cut with chemicals called restriction enzymes. **Restriction enzymes** are very specific chemicals because they always cut DNA at the same sequence of bases. They are like very specialized DNA scissors.

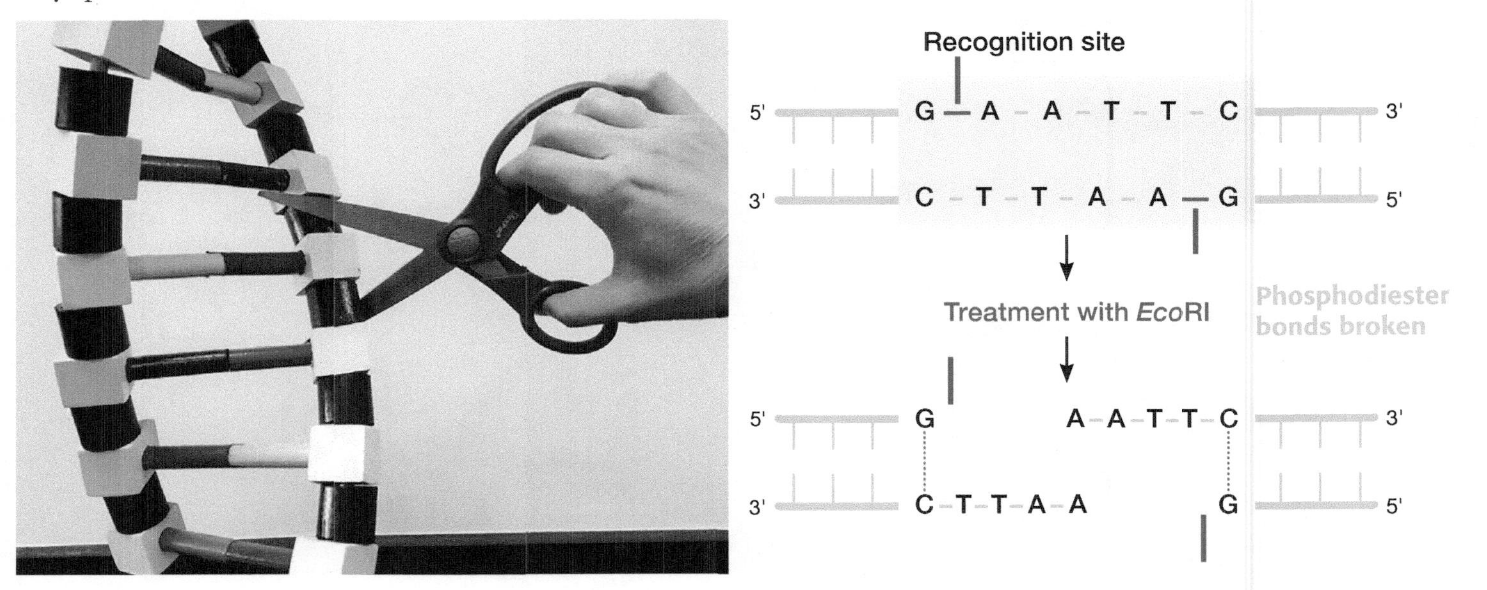

Once the fragments of DNA have been generated, they can be separated using a process called gel **electrophoresis**. As the name suggests, electricity is used to separate the pieces of DNA, and they move through a gel-like substance, either acrylamide or agarose. Because DNA is negatively charged, the pieces will move from the negative part of the gel, the anode, to the positive side, the cathode. The bigger pieces of DNA move the slowest and smaller pieces move the fastest. Here is a picture of a DNA gel with the anode and cathode shown:

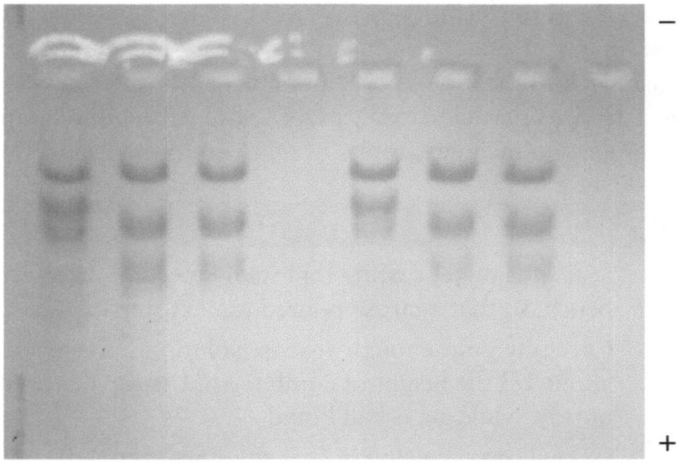

−

+

This process is useful in many different ways, and it allows us to compare DNA from two sources to see if it is from the same source. This type of analysis is used in many ways, including paternity cases, crime scene investigations and, as you will see today, in tracing foodborne outbreaks. Please discuss and list some ways which you think this type of DNA analysis may be used.

The following instructions are from a kit obtained from Carolina Biological Supply and you will run a gel on some "viral DNA" to determine if their patients are infected by a deadly virus—Good luck with your investigation!

## Outbreak! Fingerprinting Virus DNA

It is the year 2015. You are a molecular biologist working for the Centers for Disease Control and Prevention (CDC). Your job is to help track epidemics and to monitor emerging diseases.

Five years ago, a cluster of cases of hemorrhagic fever occurred in an isolated town in northeastern Alabama. The disease killed approximately 30% of those who caught it. The most alarming aspect of the new disease was that it was highly contagious from human to human, thus posing the threat of an epidemic. Medical authorities, including your office, believe the only reason the outbreak did not erupt into a major epidemic was that the Alabama town was so small and isolated. A CDC team traced the disease to a virus carried by the numerous local squirrels. An extensive trapping campaign was carried out in an attempt to eliminate the virus by eliminating the squirrels infected with it. No further cases have been reported in Alabama.

Three years ago a suspicious outbreak occurred in Pennsylvania. Several people fell ill with a hemorrhagic fever. The symptoms of the disease were the same as those of the Alabama fever, but no one died. The Pennsylvania fever was apparently less contagious than the Alabama fever since the Pennsylvania patients were exposed to many people who did not come down with the disease. However, the Pennsylvania virus was also traced to the local squirrel population.

You were asked to compare the viruses that caused the two outbreaks. You found that the virus particles looked similar. Both viruses had a DNA genome, but the base sequences of the genomes were different in many places.

Now, three people in Missouri have fallen ill with a hemorrhagic fever. Their symptoms are similar to the symptoms of the Alabama and Pennsylvania fevers. Local medical personnel were alarmed and called the CDC to determine if the dangerous Alabama virus had reappeared. The patients have been placed in quarantine. You are flown to the scene. You must immediately determine whether the Missouri patients are infected with the highly contagious and deadly Alabama virus, the Pennsylvania virus, or some other agent.

One of the techniques you decide to use is examination of viral DNA by **restriction analysis.** Wearing protective clothing, you enter the patients' isolation rooms in the Missouri hospital, carefully draw samples, place them on ice, and rush back to the biological containment laboratory at the CDC for a variety of tests. For the restriction analysis, you isolate virus particles, extract DNA, and subject the samples to restriction digestion and agarose gel electrophoresis. In your gel you include samples of DNA from the Pennsylvania and Alabama viruses for comparison.

## Procedure A: Cast Agarose Gel—this part has been done for you—proceed to part B

1. Seal ends of gel-casting tray with tape, and insert well-forming comb. Place gel-casting tray out of the way on lab bench, so that agarose poured in next step can set undisturbed.
2. Carefully pour enough agarose solution into casting tray to fill to depth of about 5 mm. Gel should cover only about 1/3 the height of comb teeth. Use a pipet tip or toothpick to move large bubbles or solid debris to sides or end of tray, while gel is still liquid.
3. Gel will become cloudy as it solidifies (about 10 min.). Do not move or jar casting tray while agarose is solidifying.
4. When agarose has set, unseal ends of casting tray. Place tray on platform of gel box, so that comb is at negative (black) end.
5. Fill box with Tris-acetate-EDTA (TAE) buffer to level that just covers entire surface of gel.
6. Gently remove comb, taking care not to rip wells.
7. Make certain that sample wells left by comb are completely submerged. If dimples are noticed around wells, slowly add buffer until they disappear.
8. The gel is now ready to load with DNA.

   *Note:* If this will be your stopping point for the lab period, cover the electrophoresis chamber to prevent gel from drying out.

## Procedure B: Load Gel

Use micropipet to load contents of each reaction tube into a separate well in gel. Use a fresh tip for each reaction tube.

1. Draw sample into pipet or gel-loading device.
2. Steady pipet over well using two hands.
3. Be careful to expel any air in micropipet tip end before loading gel. (If air bubble forms a cap over well, DNA/loading dye will flow into buffer around edges of well.)
4. Dip pipet tip through surface of buffer, position it over the well, and slowly expel the mixture. Sucrose in the loading dye weighs down the sample, causing it to sink to the bottom of the well. Be careful not to punch tip of pipet through the bottom of the gel.

## Procedure C: Electrophorese

1. Close top of electrophoresis chamber, and connect electrical leads to an approved power supply, anode to anode (black-black) and cathode to cathode (red-red). Make sure both electrodes are connected to same channel of power supply.
2. Turn power supply on and set voltage as directed by your instructor. Shortly after current is applied, loading dye can be seen moving through gel toward positive pole of electrophoresis apparatus.
3. The loading dye will eventually resolve into two bands of color. The faster-moving, purplish band is the dye bromophenol blue; the slower-moving, aqua band is xylene cyanol. Bromophenol blue migrates through gel at the same rate as a DNA fragment approximately 300 base pairs long. Xylene cyanol migrates at a rate equivalent to approximately 2000 base pairs.
4. Allow the DNA to electrophorese until the bromophenol blue band is about 2 cm from the end of the gel. Your instructor may monitor the progress of electrophoresis in your absence; in that case, omit Steps 5 and 6.
5. Turn off power supply, disconnect leads from the inputs, and remove top of electrophoresis chamber.

**6** Carefully remove casting tray, and slide gel into staining tray labeled with your group name. Take gel to your instructor for staining.

**Draw the results of your gel here:**

## QUESTIONS

1. Compare the fingerprints of the Pennsylvania, Alabama, and Missouri virus isolates. What can you conclude about the virus infecting the Missouri patients?

2. Do these new cases represent a public health emergency? Why or why not?

3. What would you tell your supervisor at the CDC?

4. Why would the restriction fragment patterns from two different viruses be different?

## Fun Fact

If the DNA in a single human cell were stretched out and laid end-to-end, it would measure approximately 6.5 feet (2 meters). The average human body contains 10 to 20 billion miles (16 to 32 billion kilometers) of DNA distributed among trillions of cells. That is the equivalent of nearly 70 trips from the earth to the sun and back.

**http://www.enotes.com/science-fact-finder/biology/how-much-dna-typical-human-cell**

## Introduction—Microarray Technology

There are approximately 20,000 genes in the human genome, and each cell contains the same set of genes. However, each of our cells looks and acts very differently. How can it be that a skin cell and a liver cell, for example, both contain the exact same genetic material, but function so differently? Part of the answer to that question lies in the concept of "gene expression." A gene is simply a section of DNA that codes for a particular protein product. To understand gene expression, we must understand what is called the central dogma of molecular biology. The central dogma tells us that DNA, through a process called transcription, is used to make another nucleic acid, RNA. Then that RNA is used to make, through translation, proteins that do much of the work in the cell. The proteins that are present in a given cell determine the function of that cell.

DNA ⟶ RNA ⟶ Protein

So, back to gene expression. A gene is said to be expressed, or "on," if the protein that it codes for is being produced. Likewise, it is not being expressed, or it is "off," if the protein it codes for is NOT being expressed. To answer the question about the liver and the skin cells, each cell is expressing different genes at different times. It is very interesting for scientists to understand gene expression; it helps us to learn about what is happening in different cells and why they act the way they do.

**Microarray** analysis is a tool that scientists are using to examine **gene expression**. It allows us to examine thousands of genes at one time. For example, we can study the genes that are turned on in a cancer cell, and compare to the genes that are turned on in a normal cell. By identifying genes that are being expressed in the cancer cell but not in the normal cell, we can further study the possible roles of those particular genes in that cancer cell. Even genes that are not highly expressed may play an important role in the cell. Genes that control the cell cycle are usually expressed at very low amounts only at particular times in the cycle. The LACK of expression of a gene may also play an important role in the cell. A protein like p53, which signals a cell with damaged DNA to die, may not be expressed by cancerous cells, and this is a dramatic change that may allow the cancer cell to continuously cycle, divide and make a tumor! Microarray technology is currently being used to determine treatment and prognoses in many types of cancers. Using your knowledge of the central dogma, what would you look for to determine if a gene is turned on in a particular cell?

Microarray analyses are conducted by placing DNA from an organism onto the surface of a small glass slide, usually the size of a microscope slide. The DNA is placed onto the slide in known locations, and a computer is used to keep track of the location of each gene. Then RNA is taken from the cells you are studying—RNA is used as an indicator of gene expression. If RNA is present—the gene is being expressed; if RNA is NOT present, the gene is not being expressed. So we take out the RNA and use it to make DNA, using an enzyme called reverse transcriptase. We can add a color at this point, so that the DNA we make, called complementary DNA or cDNA, is colored. Then we mix it with the DNA on the microscope slide, and if it is the same, it will stick. This is based on the base-pairing rules of DNA: A always binds with T and C always binds with G. Here is an example:

A T G C G T A G A C    cDNA
**T A C G C A T C T G**    **DNA**

By labeling the cDNA from the cancer cells one color (usually red) and the cDNA from healthy cells another color (usually green), we can easily (with the help of a computer again!!) visualize the difference.

This technology can be used to study many different scenarios, not just cancer vs. non-cancer cells. Watch this animation to learn how this technology can be used to study gene expression in yeast cells grown in two different environments, one with oxygen and one without: http://www.bio.davidson.edu/courses/genomics/chip/chip.html. Another very helpful website to learn about microarray technology is http://microarray.cisat.jmu.edu/.

So, here is a summary of the steps involved in this process:

STEP 1: Using a robot, DNA is spotted onto the glass slide. The DNA printed onto a slide is oligonucleotides (bases in a known sequence order). Scientists print thousands of microscopic DNA spots on glass slides, one spot for each gene in the genome.

STEP 2: Isolate mRNA from two different cell populations, for example, normal lung and cancerous lung cells. We need a special enzyme to change mRNA to cDNA since RNA is so chemically unstable we use an enzyme called reverse transcriptase (RT) to accomplish this step.

STEP 3: As part of the RT reaction, the normal tissue cDNA is labeled with a green dye, the cancerous cell cDNA with a red dye.

STEP 4: Hybridize your microarray with labeled cDNAs from normal lung and lung cancer tissue.

STEP 5: Wash away any cDNAs that did not bind and ready the chip for laser reading!

STEP 6: Visualize your labeled microarray results; in immunology we ship our chips to NC for scanning with lasers that make the dots glow red, green or orange! Then the real fun starts: data analysis of 30,000 (or more!) spots!

In this exercise, you will use a simulated microarray kit to illustrate this principle and to learn how this technique can be used in cancer research. You will use a simulated DNA microarray ("gene chip") to study the expression of six different genes in healthy lung cells and lung cancer cells. Prior to beginning this experiment, look up each gene and give a brief (1 or 2 sentence) description of its function. Include where this gene is normally expressed (liver, lung, blood, etc.). It is also important to note if this gene has been designated as a "housekeeping gene," "oncogene," "tumor suppressor" or if this gene is involved in apoptosis (programmed cell death)! These are broad categories of genes; your instructor will discuss the differences with you before you begin.

GENE 1: MYC—Myelocytomatosis

GENE 2: GAPDH—Glyceraldehyde phosphate dehydrogenase

GENE 3: SFTPD—Surfactant associated protein

GENE 4: HBG1—Hemoglobin, gamma A

GENE 5: BCL2—B cell, CLL2

GENE 6: MUC1—Mucin 1

For each gene, hypothesize whether you think it will be expressed in the normal cell, the cancer cell, in both or in neither. Write yes or no in the following table as appropriate.

| | Gene will be expressed in: | | | |
|---|---|---|---|---|
| | Cancer cell | Normal cell | Both cell types | Neither |
| MYC | | | | |
| GAPDH | | | | |
| SFTPD | | | | |
| HBG1 | | | | |
| BCL2 | | | | |
| MUC1 | | | | |

## Procedure

**WARNING:** Hybridization Solution contains 0.4M Sodium Hydroxide (NaOH), which is toxic. It can cause skin burns. Do not get in your eyes, on skin, or on clothing. ***Wear gloves and use goggles.*** **FIRST AID:** In case of contact: flush with water. **DO NOT DRINK, but if you ingest or have eye contact, flush with copious amounts of water and seek medical attention!**

Time to print your "DNA" onto a glass slide. In real DNA microarrays, scientists also print thousands of microscopic DNA spots on glass slides, one spot for each gene in the genome. Your gene chip is a macroarray: your spots will be much larger and you will be able to view them without specialized equipment.

1. For this lung cancer DNA kit, the six genes that you will study are those you researched listed on page 89.
2. Do not touch the surface of your slide (handle only on the edges). Using the permanent marking pen, label the red backgrounds next to the 6 spots furthest from the label end with numbers 1–6, making two rows of three. Write something to identify your group on the label end of the slide (initials, 831, etc.).
3. Bring your labeled slide to the water bath area.
4. Using the dropper bottle, carefully spot the appropriate gene solutions into each of the labeled areas of your slide. Be sure to place the correct DNA sequence in the correct spot. Once the spots are hardened and dry, your microarray has been successfully "printed" with the 6 genes!

   **NOTE:** The mRNA was isolated from normal lung cells and its cDNA was labeled with a blue dye. mRNA from lung cancer cells was isolated and its cDNA was labeled with a pink dye. You will not be able to see these dyes until you visualize your results at the end of the experiment because they are very diffuse.
5. Wearing gloves, carefully drop 2–3 drops of "Hybridization Solution" from the dropper bottle onto each spot. **DO NOT allow the dropper bottle to touch the DNA spots!**

   **NOTE:** The bottle (to be shared between groups) contains a solution of labeled cDNAs from lung cancer cells and normal cells mixed together. You cannot see the color because the cDNA is very dilute. When added to your printed slide, the labeled cDNAs in the solution will base pair with the complementary DNA for each gene spotted onto the macroarray. This is the principle behind real DNA microarrays. Binding cDNA to spotted DNA will produce a visible color, so watch carefully!
6. Once you see some color appear, place it on a sheet of white paper and record your results below. You may be able to slow the color development if it gets too dark by dunking the slide in a beaker of water.

## ANALYSIS OF RESULTS

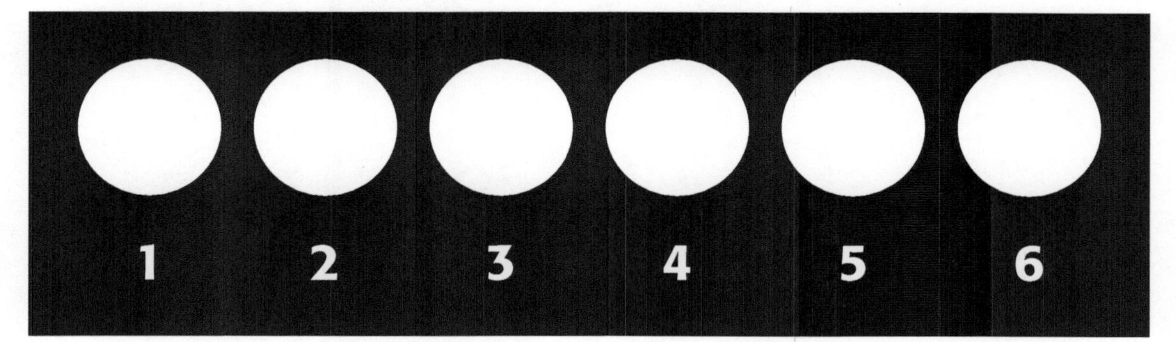


1. Which gene(s) were expressed (transcribed) in the lung cancer cells? How do you know?

2. Which gene(s) were not expressed in the lung cancer cells? How do you know?

3. Were there any genes not expressed in either cell type? Explain what kind of gene this could be.

4. Were there any genes expressed in both cell types? Explain what kind of gene this could be.

5. Which genes may play a role in causing cancer in lung cells? Explain why you chose these genes and not other genes.

6. Work with your group to discuss another microarray experiment that you would like to carry out. Be sure to include how you will set up your experiment, what cells you will use, etc. It cannot involve cancerous cells! Be prepared to discuss this with the class.

# Review Questions

**1** Draw a molecule of DNA showing complementary base pairs.

**2** What does it mean for a gene to be expressed?

**3** What is the central dogma of molecular biology?

**4** You have isolated 6 fragments of DNA—their sizes are shown below. Please draw how they would appear on an electrophoretic gel. You ran a size ladder along with your DNA—those bands are already shown on the gel and the sizes were 500, 1000, 2500, 5000, 7500 kbp. Please label them as well.

Fragment 1—500 kbp
Fragment 2—4000 kbp
Fragment 3—2700 kbp
Fragment 4—250 kbp
Fragment 5—6500 kbp
Fragment 6—3750 kbp

**5** List three ways that DNA technology is used today.

**Optional homework:** DNA technology has allowed us to modify many organisms, and to actually put the DNA from one organism into the DNA of an entirely different species. We are using this technology to modify both plants and animals used to produce our food. Please find an example of a genetically modified organism used for food, and discuss your opinion of that process in a one-page (single-spaced) paper.

# Lab Practical

Enzymes
Mitosis and the Cell Cycle
Genetics/Fast Plants
DNA Fingerprinting
Microarray

# This Fine Mess
## Understanding Protists

# 9

*Although traditional classifications of protists tended to have little respect for phylogeny—neither in theory nor the data were available to make this possible—they did have one great advantage. They were designed to be user-friendly.*

—Colin Tudge (1943–present)

## OBJECTIVES

*At the completion of this chapter, the student will be able to:*

1. Describe the general characteristics of the protists.
2. Discuss why protists present a unique problem for taxonomists.
3. Compare and contrast traditional and modern protist classification.
4. Explain the characteristics of representative plantlike, funguslike, and animallike protists.
5. Recognize various protists using the microscope.
6. Discuss the roles of protists in the environment and in medicine.

*Answer the following questions prior to coming to lab.*

1. What are several important medical and commercial uses of algae?

2. How did a simple protist change the history of Ireland?

3. What is the importance of a protist known as *Globigerina*?

4. In the new classification scheme, what are the basal protists and their fundamental characteristics?

5. Did a protist change the life of Charles Darwin?

# Protists

Perhaps the British biologist Colin Tudge said it best when trying to make sense of the hodgepodge kingdom known as Protista when he declared, "Although traditional classifications of protists tended to have little respect for phylogeny—neither in theory nor the data were available to make this possible—they did have one great advantage. They were designed to be user-friendly." The 100,000 named members of kingdom Protists are a polyphyletic group of organisms underlying the kingdoms Plantae, Fungi, and Animalia. This diversity is the main reason the kingdom Protista is no longer considered viable, though it is still a convenient, and widely understood, term to describe this diverse group.

Classically, the protists have been categorized as plantlike protists, fungus-like protists, and animallike protists in accordance with their roles in the environment. This scheme, although antiquated, is still a useful tool in developing a basic understanding of the protists. Recently, studies using DNA sequencing and cytological analysis have painted a more complex picture of the protists. This new look at the protists suggests the once singular kingdom should be divided into several supergroups that can be divided further into kingdoms (as many as 30, according to some authors). (Table 9.1 on page 108 describes this further.) As time passes, a clearer picture of protist classification will emerge.

The term *protist* describes microscopic, unicellular organisms, such as the amoeba, much of the algae in your ditch, and gigantic multicellular organisms exceeding 600 feet in length, such as kelp. Protists are ancient eukaryotes that first appeared in the Precambrian era nearly 2 billion years ago. A multitude of protists can live in a drop of water, in the terrestrial environment, or in the body of a host. In the environment, protists can function as autotrophs, heterotrophs, and even decomposers. Several protists are responsible for diabolical parasitic infections in humans and other animals.

Recall Exercise 3.3 in Chapter 3 when you made a wet mount of pond water and how you were amazed at some of Leeuwenhoek's wee beasties dashing and swimming about in a single drop of water? Many of these organisms from the algae to the acrobatic paramecia were protists.

# Plantlike Protists

The plantlike protists, commonly called **algae**, are extremely diverse, ranging from minute diatoms to giant, multicellular kelp. These organisms exist as photoautotrophs possessing chlorophyll *a* and membrane-bound plastids. They produce copious oxygen and serve as the basis for the food chain. Traditionally, the names of the phyla were derived based upon accessory pigmentation. In this exercise, we will conduct procedures on specimens of phyla Chlorophyta, Bacillariophyta, and Euglenophyta. Phyla not included in this exercise but still considered protists are Rhodophyta (red algae such as nori used to wrap sushi and *Agar* spp. used as a growing medium for bacteria) Phaeophyta (brown algae such as kelp and *Sargassm*) and Pyrrophyta/Dinophyta (fire algae such as *Gonyaulax catenella*, also considered an animallike protist, that causes red tide).

## Plantlike Protists

### Phylum Chlorophyta

Phylum **Chlorophyta**, known as green algae, comprises about 10,000 unicellular and multicellular species (Fig. 9.1). Most green algae are aquatic, but some species can be found growing on tree trunks, sidewalks, buildings, unwashed cars, and even snow. One species even lives on the fur of tree sloths. Green algae are thought to be the ancestors of modern plants. However, not all green algae are green in color. We will look at two examples of chlorophytes: *Spirogyra* spp. and *Volvox* spp.

*Spirogyra crassa* is a classic example of this phylum. Sometimes called watersilk, it is a filamentous freshwater species that possesses ornate spiral chloroplasts (Fig. 9.2). It can be found floating in mats on the surface of quiet ditches and ponds. *Spirogyra* is capable of a type of sexual reproduction known as **conjugation**. During this process, a conjugation tube forms between adjacent filaments. **Protoplasts** (protoplasm of cell with cell wall removed) from one filament migrate via the tube to another filament to fuse with a waiting protoplast. The mobile protoplast is considered male, and the stationary protoplast is considered female. A thick-walled zygote results. Eventually the zygote undergoes meiosis, forming four haploid cells. Three disintegrate, and the remaining cell forms a new *Spirogyra* filament.

*Volvox* is a common colonial freshwater alga found in ponds, ditches, and puddles (Fig. 9.3). The colony is composed of as many as 50,000 individual flagellated cells that form a hollow sphere. It often resembles a basketball with its cells being homologous to pebbles. Each cell has an **eyespot** that helps locate the light necessary for photosynthesis. Some colonies are asexual, composed of nonreproducing vegetative cells and gonidia that produce new **daughter colonies**. *Volvox* is capable of sexual reproduction, with male colonies producing and releasing microgametes and female colonies producing female sex cells, or macrogametes.

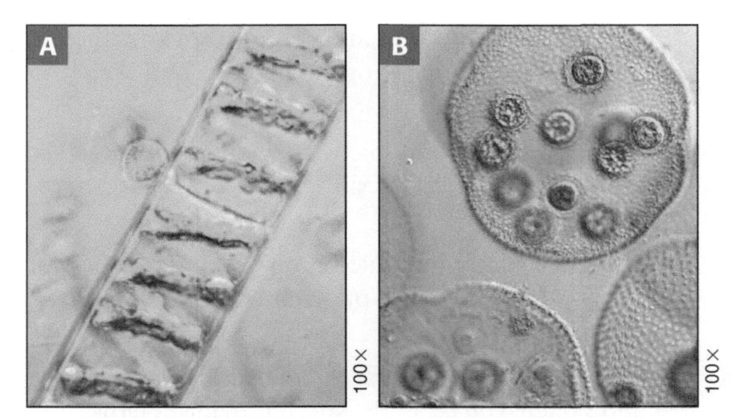


FIGURE **9.1** Examples from the phylum Chlorophyta: (**A**) *Spirogyra* sp.; (**B**) *Volvox* sp.

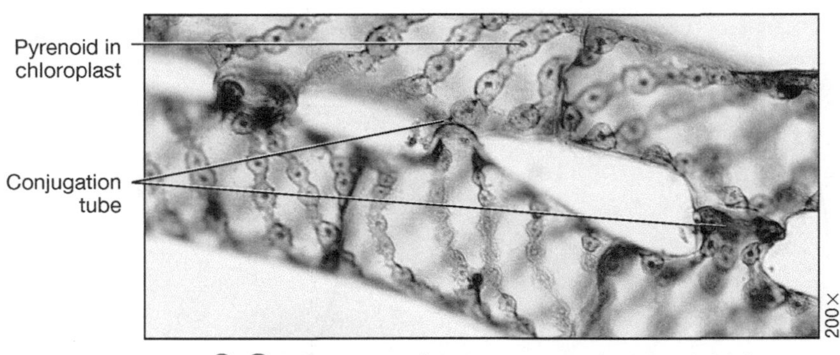


FIGURE **9.2** Filaments of *Spirogyra* sp. showing initial contact of conjugation tubes.

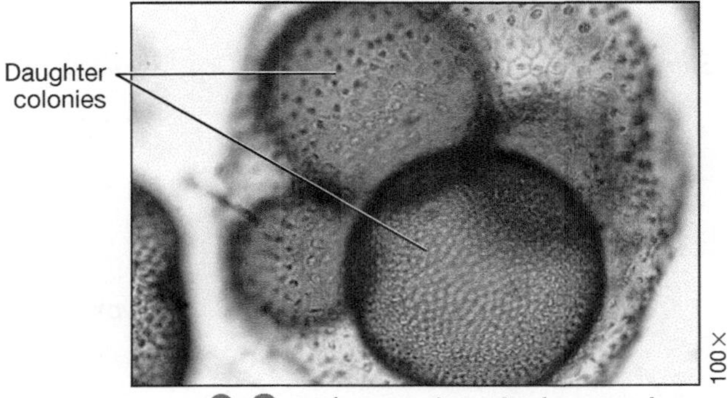


FIGURE **9.3** *Volvox* sp. is a single organism with several large daughter colonies.

## Procedure 1

### Observing *Spirogyra* and *Volvox*

**Materials**

- ❑ Compound microscope
- ❑ Prepared slides of *Spirogyra* and conjugating *Spirogyra*
- ❑ Prepared slides of *Volvox*
- ❑ Herbarium sheets
- ❑ Wet mounts
- ❑ Microscope slides and coverslips
- ❑ Eyedropper
- ❑ Colored pencils
- ❑ Culture of *Spirogyra*
- ❑ Culture of *Volvox*

1. Procure the equipment and specimens. Using the compound microscope and a prepared slide of *Spirogyra*, describe and sketch *Spirogyra* in the space provided below.
2. Using the compound microscope and a prepared slide of *Spirogyra* undergoing conjugation, describe and sketch *Spirogyra* conjugation in the space provided below.
3. Using the compound microscope, prepare a wet mount of *Spirogyra* from the provided culture. Note your observations and sketches in the space provided below.
4. Using the compound microscope and a prepared slide of *Volvox*, describe and sketch *Volvox* in the space provided below.
5. Using the compound microscope, prepare a wet mount of *Volvox* from the provided culture. Note your observations and sketches in the space provided below.

*Spirogyra* Magnification ____________

*Spirogyra* Magnification ____________

*Spirogyra* Magnification ____________

*Volvox* Magnification ____________

*Volvox* Magnification ____________

## Phylum Bacillariophyta

Phylum **Bacillariophyta** consists of uniquely shaped algae called **diatoms** that possess a **test,** or shell, made up of two halves. These organisms are the most numerous unicellular algae found in marine and freshwater environments. Millions of diatoms may exist in a liter of water. There are more than 10,000 species, and some biologists estimate a million species. Diatoms are exquisite organisms, appearing in a variety of geometrical shapes (Fig. 9.4). A large component of the test is silica ($SiO_2$), a major component of glass incorporated into an organic mesh. Because their test is composed of silica, the fossil record of diatoms is well represented. They make up diatomaceous earth used in silverware polish, insulation, swimming pool filters, reflective paint, and even toothpaste. Some deposits of diatoms are more than 3,000 feet thick. A common diatom is *Cyclotella stelligera*.

## Procedure 2
### Observing Diatoms

**Materials**
- ❑ Compound microscope
- ❑ Prepared slides of diatoms
- ❑ Microscope slides and coverslips
- ❑ Eyedropper
- ❑ Colored pencils
- ❑ Culture of diatoms

1. Procure the equipment and specimens.
2. Prepare a wet mount of diatoms from the culture provided. Using the compound microscope, compare your slide with the prepared slide of diatoms. Note your observations and sketches in the space provided on the following page.

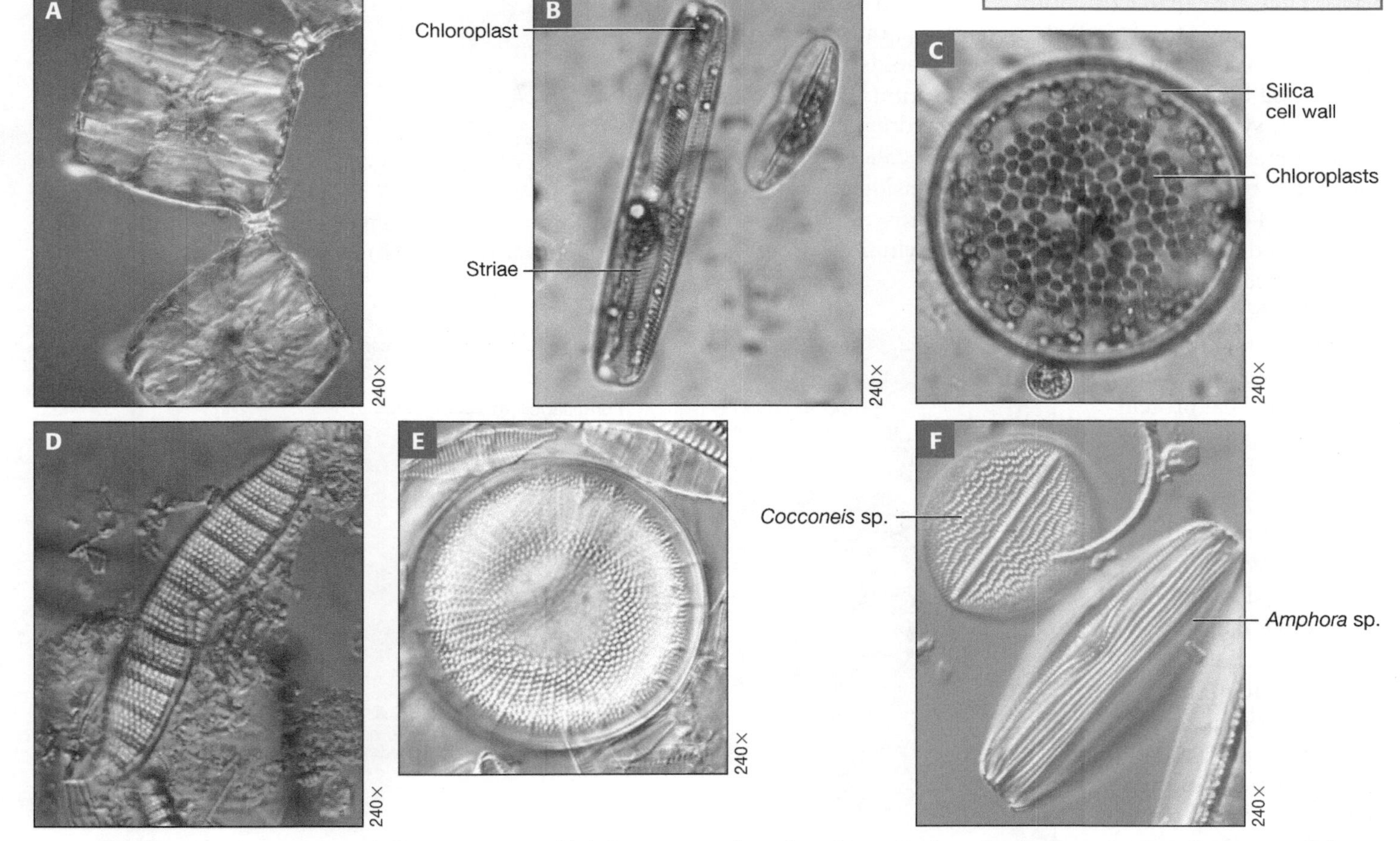


FIGURE **9.4** Several examples of diatoms: (**A**) *Biddulphia* sp., a colony forming colonies; (**B**) live specimens of pennate (bilaterally symmetrical) diatoms: *Navicula* sp. (left), and *Cymbella* sp. (right); (**C**) *Hyalodiscus* sp., a centric (radially symmetrical) diatom from a freshwater spring in Nevada; (**D**) *Epithemia* sp., a distinctive pennate freshwater diatom; (**E**) *Stephanodiscus* sp., a centric diatom; (**F**) two common freshwater diatoms.

Specimen ______________________

Magnification ______________________

______________________

______________________

______________________

Specimen ______________________

Magnification ______________________

______________________

______________________

______________________

## Phylum Euglenophyta

Phylum **Euglenophyta** includes approximately 1,000 species of unicellular, flagellated, freshwater species. Nearly 40 genera of euglenoids have been described (Fig. 9.5). One-third of these genera possess chloroplasts, and the other two-thirds do not have chloroplasts. They are capable of photosynthesis and can capture their own food—which presents a paradox for taxonomists. Nonphotosynthetic euglenoids gather food by particle feeding and absorption. These euglenoids synthesize a starch-like carbohydrate called paramylon, stored in conspicuous particles called paramylon bodies. Some euglenoids are considered **mixotrophs** because they can undergo photosynthesis and can feed. A common species of this phylum is *Euglena deses*.

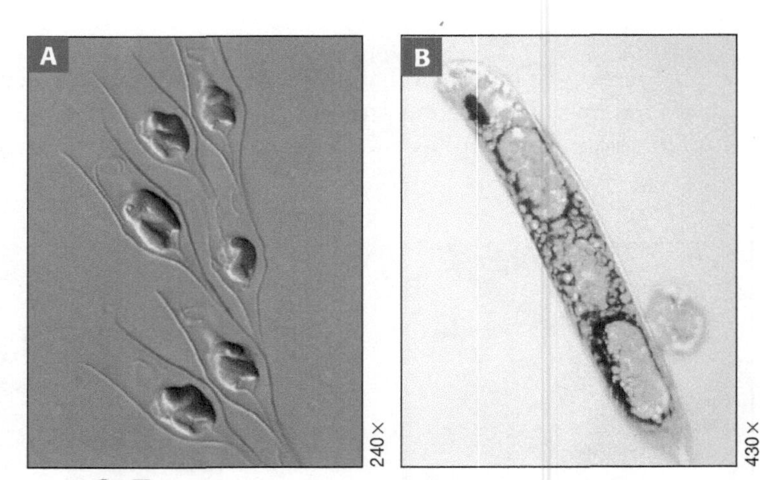


FIGURE **9.5** Examples from the phylum Euglenophyta: (**A**) *Dinobryon divergens* sp.; (**B**) *Euglena* sp.

Members of the genus *Euglena* are characterized by having a **pellicle** (helical protein bands that extend along the length of the cell beneath the plasma membrane) rather than a rigid cell wall. In addition, individuals possess two **flagella** (a short flagellum and a long flagellum) used for locomotion; a paramylon body; a gullet, through which food can be ingested; and a red stigma, or eyespot, that aids in light detection (Fig. 9.6).

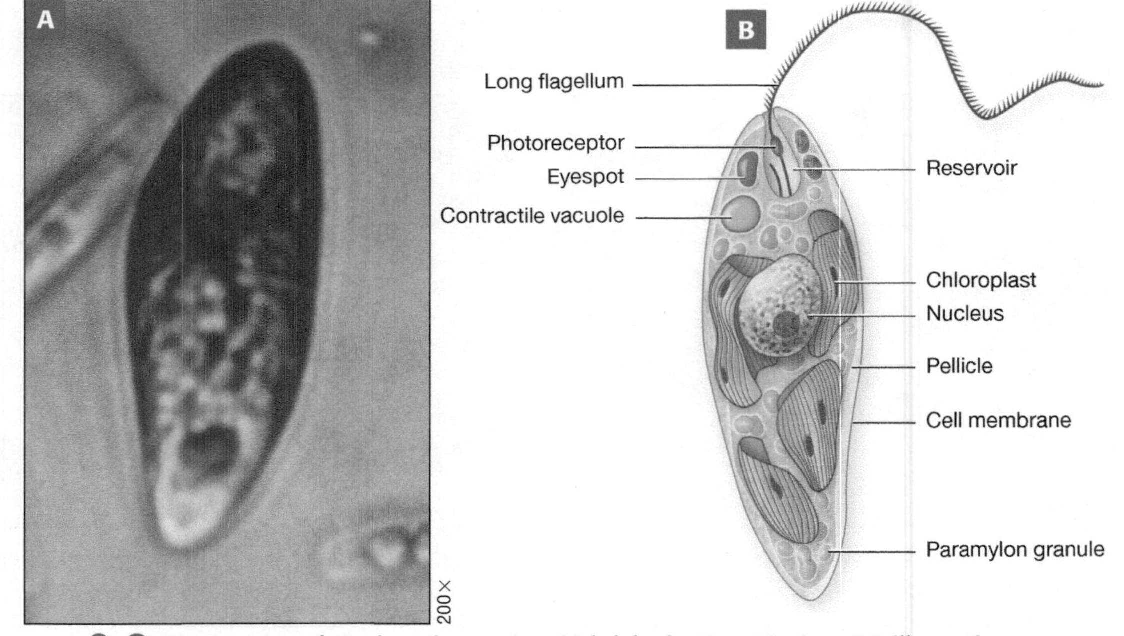


FIGURE **9.6** (**A**) Species of *Euglena* from a brackish lake in New Mexico; (**B**) illustration.

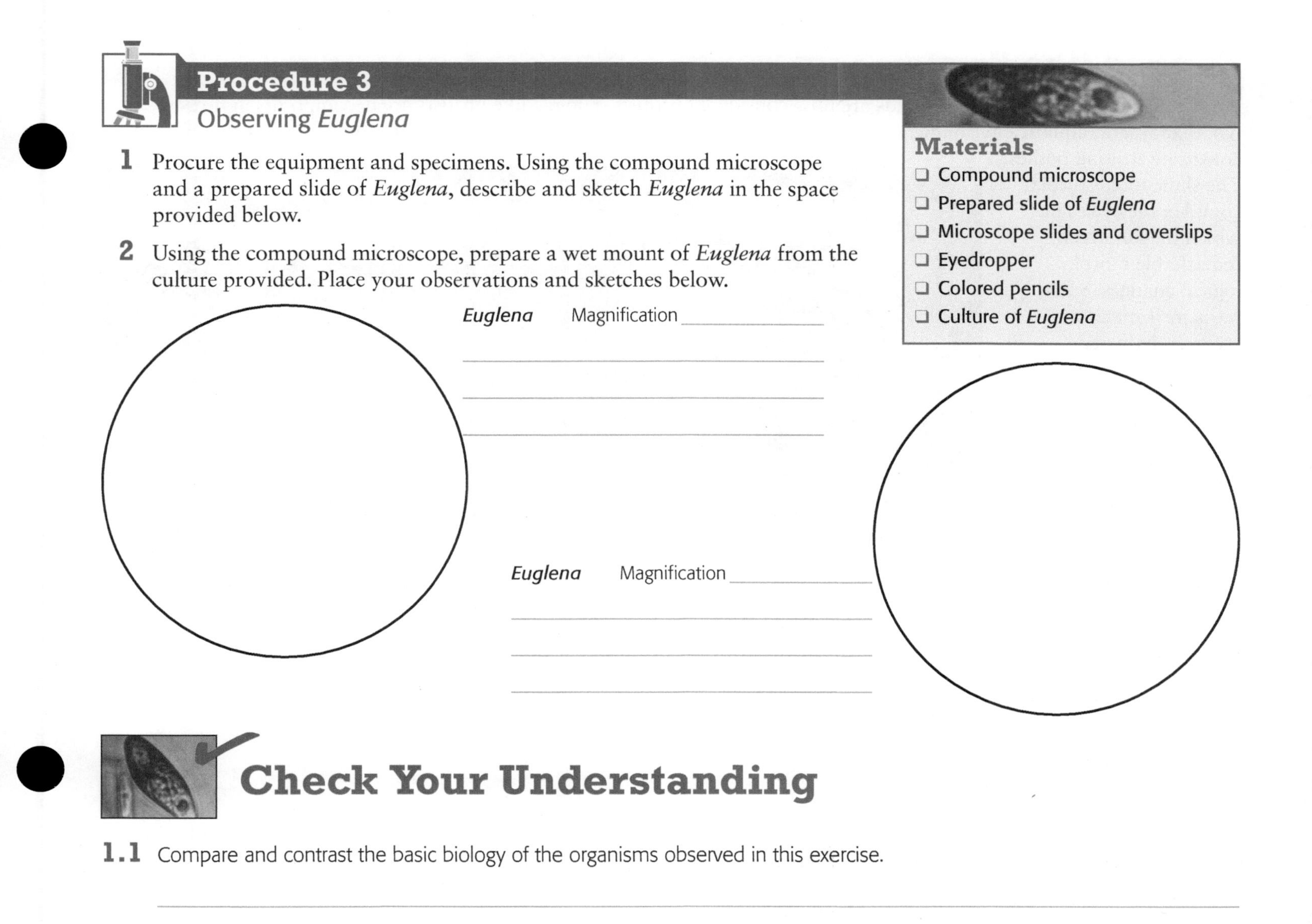

## Procedure 3
### Observing *Euglena*

**Materials**
- ❑ Compound microscope
- ❑ Prepared slide of *Euglena*
- ❑ Microscope slides and coverslips
- ❑ Eyedropper
- ❑ Colored pencils
- ❑ Culture of *Euglena*

1. Procure the equipment and specimens. Using the compound microscope and a prepared slide of *Euglena*, describe and sketch *Euglena* in the space provided below.
2. Using the compound microscope, prepare a wet mount of *Euglena* from the culture provided. Place your observations and sketches below.

*Euglena* Magnification ______________

______________________________

______________________________

______________________________

*Euglena* Magnification ______________

______________________________

______________________________

______________________________

## Check Your Understanding

**1.1** Compare and contrast the basic biology of the organisms observed in this exercise.

______________________________

______________________________

______________________________

**1.2** Describe the movement of *Volvox* and *Euglena*.

______________________________

______________________________

**1.3** Why is *Euglena* called a mixotroph?

______________________________

______________________________

**1.4** Describe the form of diatoms.

______________________________

______________________________

# Fungus-like Protists

Once considered fungi, the **slime molds** are now considered protists. Fungus-like protists are decomposers in forests, woodlands, and aquatic environments. Slime molds are motile organisms and produce spores on sporangia that in turn constitute fruiting bodies. The slime molds display complex life cycles in which they undergo remarkable morphological changes. Some measure 1 meter or more in diameter (Fig. 9.7). There are cellular slime molds and plasmodial slime molds. You will look at plasmodial slime molds in this exercise.

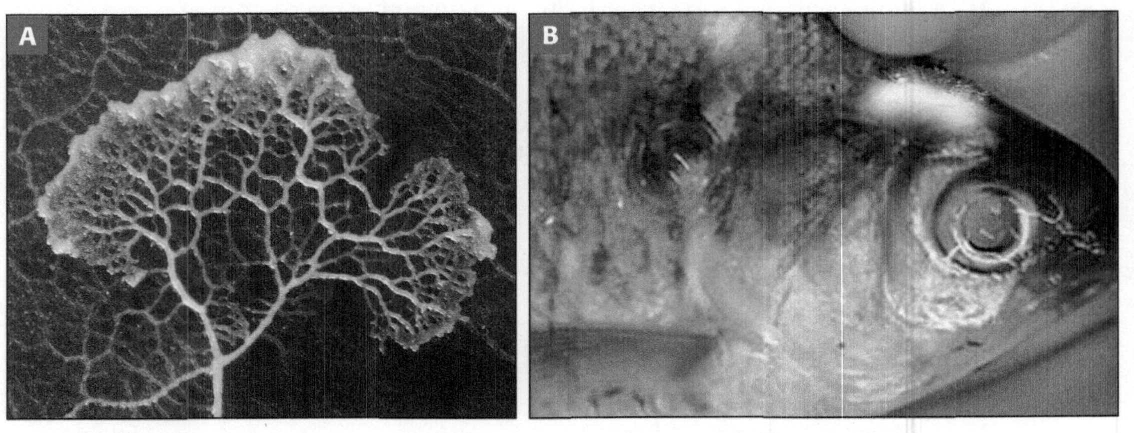


FIGURE **9.7** Fungus-like protists: (**A**) *Physarum* sp., a slime mold; (**B**) *Saprolegnia* sp., a water mold.

## Fungus-like Protists

### Phyla Acrasiomycota and Oomycota

The **cellular slime molds** have been placed in phylum **Acrasiomycota.** These organisms exist as solitary amoeboid cells in the soil. During adverse conditions, they aggregate, forming a mass called a pseudoplasmodium. This temporary stage gives rise to fruiting bodies that produce spores.

**Water molds** are placed in phylum **Oomycota**. One water mold, *Saprolegnia* spp., can be easily observed as a cotton-like filamentous mass growing on dying or dead fish. The water molds have a cell wall composed of cellulose, unlike the fungi, which have a cell wall composed of chitin. One species, *Phytophthora infestans*, was responsible for the Irish potato famine in the 1840s. This organism had a direct impact on U.S. history because it forced a large number of Irish people to immigrate to the United States.

### Phylum Myxomycota

Phylum **Myxomycota** consists of about 500 species known as the **plasmodial slime molds.** These organisms exist as a multinucleate mass covered in a sheath of slime. Many are colored bright yellow or orange and resemble a giant amoeba sliming ever so slowly across the forest floor. *Physarum* spp. is a common slime mold found in the woodlands of North America.

In this exercise, you will grow the plasmodium slime mold *Physarum polycephalum* on a petri dish with agar (growth media) and oatmeal flakes. Over a period of a week, you will be able to observe the active plasmodial stage of the slime mold, including protoplasmic streaming.

When nutrients (growth media/oatmeal flakes) are no longer available, the slime mold will enter into the sclerotia stage. This is the inactive, or dormant, stage, and the mold can remain in this state for years. If nutrients and proper environmental conditions recur, fruiting bodies form within 12 hours, and sporulation will occur.

## Procedure 1

### Observing Plasmodial Slime Mold

**Materials**

- ❑ Sterile petri dish containing nutrient agar
- ❑ Sterile oatmeal flakes
- ❑ Culture of *Physarum polycephalum*
- ❑ Sterile forceps
- ❑ Sterile scalpel
- ❑ Marker for labeling petri dish
- ❑ Dissecting microscope
- ❑ Colored pencils
- ❑ Wax pencil

**1** Procure the materials from your instructor. Using a sterile scalpel, carefully cut a tiny cube of the *Physarum* culture from the stock dish and place it upside down on your sterile petri dish containing agar. Using the sterile forceps, place two oatmeal flakes 1 inch away from the *Physarum* culture. Place the lid back onto the dish, and with a wax pencil write your name/class section on a corner of the lid.

**2** Each day, add two or three more oatmeal flakes, following the same procedure as in step 1. Observe and record the growth and streaming of the plasmodium. Describe the growth of the slime mold over a period of a week. What color is the plasmodium? Describe the morphology of the slime mold.

______________________________________________

______________________________________________

______________________________________________

______________________________________________

______________________________________________

______________________________________________

**3** After viewing it with a dissecting microscope, describe the macroanatomy of your specimen. Record your observations in the space next to each specimen.

**4** Clean up and dispose of your materials according to the instructor.

| *Physarum* Magnification ________ | *Physarum* Magnification ________ | *Physarum* Magnification ________ |
|---|---|---|

## Did you know . . .

### Did You Say Dog Vomit?

The plasmodium slime mold *Fuligo septic* is placed in the order Myxomycota. Because it can appear almost overnight and appears as a slimy mass, it gets the name dog vomit slime mold, although it is bright yellow or orange in appearance. *Fuligo* can often be found living on bark mulch in urban areas after a heavy rain or excessive watering. It also readily grows on rotten wood and plant litter. *Fuligo* has a life cycle similar to other plasmodial slime molds. Despite its very unpleasant name, *Fuligo* is completely harmless to humans, animals, and plants. Only in rare cases it may be linked to asthma problems.

## Check Your Understanding

**2.1** Where can one find slime molds?

**2.2** What was the organism responsible for the Irish potato famine?

**2.3** Where can one find *Saprolegnia*?

## Did you know . . .

### Crash Landing

Female mosquitoes are the blood feeders, while the male mosquitoes eat pollen. The male mosquito can be distinguished by a more elaborate antenna system.

One can distinguish between the *Anopheles* spp. mosquito and other species of mosquitoes by the way they land. The *Anopheles* mosquito holds its abdomen at a 45-degree angle to the host, whereas other species keep their abdomen parallel to the host. Knowing this, don't give the "guys" a break, because it takes two to make more mosquitoes. Also, don't give any mosquito a break by the way it lands. Other species carry other diseases including the West Nile virus.

Female mosquito feeding on blood.

# Animallike Protists

Remember the first time you examined ditch water with a microscope? Wow! It was like viewing a miniature version of *Star Wars* with all of the tiny organisms gliding around in a thin film of water. Antony van Leeuwenhoek shared your fascination centuries before, terming these organisms "animacules and cavorting wee beasties." Today we know that many of Leeuwenhoek's wee beasties actually were animallike protists sometimes called protozoans (Fig. 9.8). Classically, the animallike protists are divided into four major groups:

1. those that move by false feet
2. those that move by flagella
3. those that move by cilia
4. nonmotile forms

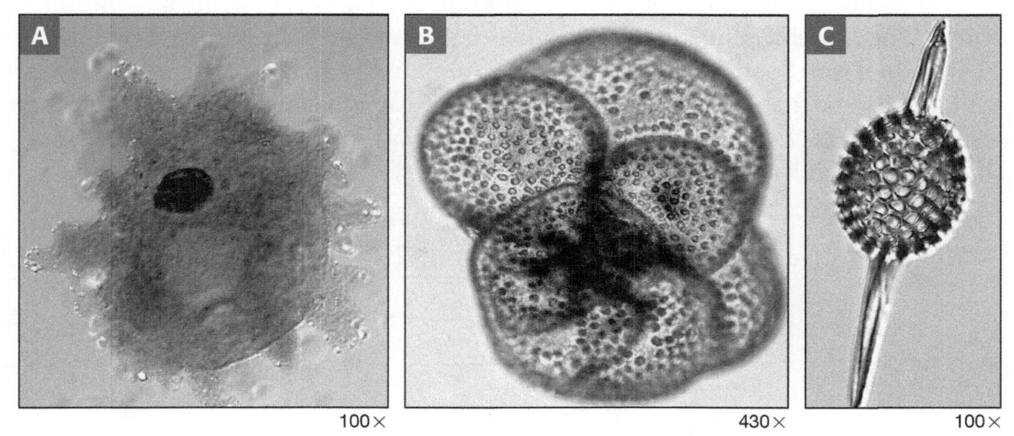


FIGURE 9.8 Animallike protists: (A) *Amoeba proteus*; (B) *Globigerina* sp.; (C) *Stylatractus* sp.

In the following exercise, we will look at the first three types of animallike protists. The fourth group, nonmotile forms, are placed in phylum Apicomplexa. These organisms exist as endoparasites in a number of hosts. Apicomplexans lack locomotor structures in the adult form. The complex life cycle usually has a sexual phase and an asexual phase. In many species, the life cycle involves multiple hosts. The best-known apicomplexan (*Plasmodium* spp.) is responsible for causing malaria ("bad air"). Between 350 and 500 million clinical episodes of malaria occur yearly, resulting in at least 1 million deaths. Malaria is the leading cause of death in children under 5 years of age. The vector for malaria is the female *Anopheles* spp. mosquito. Malaria has been a formidable disease throughout human history and presently seems prime for a comeback, considering the ease of world travel. Another apicomplexan of note is *Toxoplasma gondii*, responsible for causing toxoplasmosis in humans. The Centers for Disease Control and Prevention (CDC) estimates that perhaps one-third of the world's human population carries this parasite. Although it infects a variety of mammals, cats are the primary host. Healthy adults rarely contract the disorder; individuals with a weakened immune system and fetuses are at greatest risk. Toxoplasmosis can cause severe nerve damage, hydrocephaly, and death in newborns. Therefore, pregnant women should bring their cats to the veterinarian for a checkup and avoid changing litter boxes and gardening in areas where cat feces are common.

## Animallike Protists

### Amoeba and Amoeba-like Protists

Phylum Sarcodina (Rhizopoda)

The animallike protist called the amoeba was first described in 1757 by German naturalist August Johann Rösel von Rosenhof (1705–1759). *Amoeba proteus* is a relatively common protist that inhabits aquatic environments. Early naturalists called the amoeba *Proteus animalcule* after the Greek god Proteus, who could change his shape. The ability of some of the animallike protists to move via false feet, or **pseudopodia** (used for locomotion and for capturing food), makes them the ultimate shape-shifters of the living world. In the amoebae, pseudopods are lobe-shaped and tipped by a **hyaline cap**, a clear space at the leading edge of the pseudopod.

Members of this phylum are found primarily in freshwater and marine environments, although several parasitic species exist. Classically, amoeboid protists, such as the common freshwater amoeba (*Amoeba proteus*), have been placed in phylum **Sarcodina** (**Rhizopoda**). The sarcodines do not have a wall, or pellicle, surrounding their plasma membrane. The majority of sarcodines reproduce asexually through fission. One species, *Entamoeba histolytica*, is an intestinal parasite responsible for amoebic dysentery spread in contaminated food.

In addition to locomotion, the pseudopodia are important in helping the amoeba surround its food through **phagocytosis.** The amoeba is surrounded by a plasma membrane that possesses two distinct parts: the outer **ectoplasm** and an inner **endoplasm. Food vacuoles** are also apparent in the amoeba (Fig. 9.9). They serve to hold food particles and fuse with a lysosome that aids in digesting the food. Waste is usually eliminated by exocytosis. The nucleus, or in some species several nuclei, is easily seen in amoebae. Many times the smaller nucleolus is also apparent within the nucleus. Occasionally, an osmoregulatory **contractile vacuole** can be seen within an amoeba.

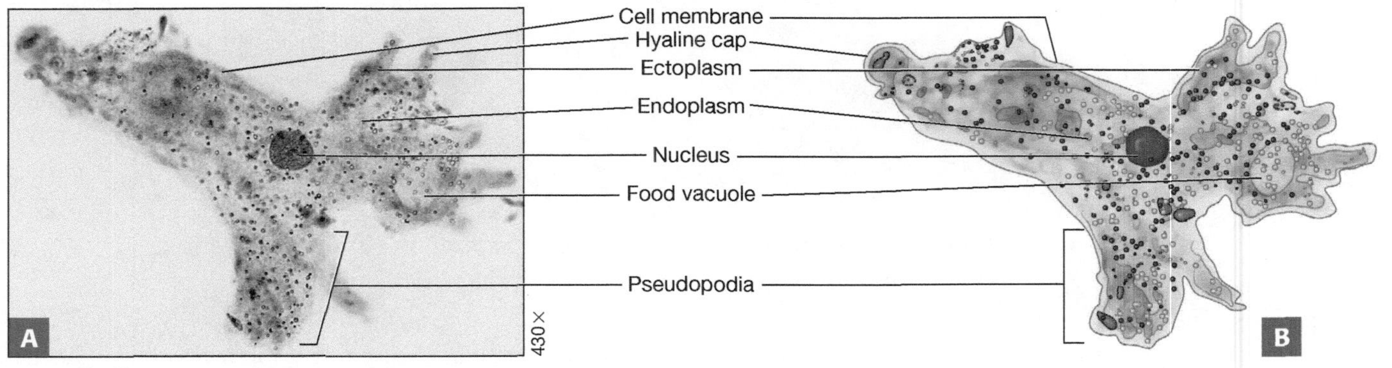


FIGURE **9.9** (**A**) *Amoeba proteus;* (**B**) illustration.

### Phylum Foraminifera

**Phylum Foraminifera** is an intriguing group of aquatic and marine amoeboid protists with elaborate, colorful shells composed of calcium carbonate. These organisms have threadlike, branched pseudopods that extrude from pores in the shell. Foraminiferans are found in tremendous numbers in the oceans. It is estimated about one-third of the sea floor consists of the shells, or casts, from dead foraminiferans, such as *Globigerina* sp. Fossil foraminiferans, or forams, are responsible for forming limestone. The white cliffs of Dover are composed primarily of the remains of foraminiferans.

### Phylum Actinopoda

The radiolarians (**phylum Actinopoda**) possess an endoskeleton composed of silicon dioxide. Radiolarians, with their thin, relatively stiff pseudopods, are found in the marine environment. Many consider the radiolarians, with their geometric endoskeletons, the most beautiful organisms on Earth. Some radiolarians, such as *Lychnaspis miranda*, resemble snowflakes. In some parts of the ocean, radiolarian sediment on the floor can be more than 4,000 meters thick.

## Procedure 1

### Observing Amoebas and Amoeba-like Protists

**Materials**
- ❑ Compound microscope
- ❑ Prepared slide of *Amoeba* sp.
- ❑ Prepared slide of foraminiferans
- ❑ Prepared slide of radiolarians
- ❑ Microscope slides and coverslips
- ❑ Eyedropper
- ❑ Culture of *Amoeba* sp.
- ❑ Colored pencils

### Procedure 1

In this procedure you will observe and record the basic anatomy of *Amoeba* sp., foraminiferans, and radiolarians (Fig. 9.10).

1 Procure a slide of the provided species of amoeba, and record which species is used in the space provided. Observe the specimen using low and high power. Locate the pseudopods, plasma membrane, hyaline cap, ectoplasm, endoplasm, nucleus, nucleolus, food vacuoles, and contractile vacuole. Describe, sketch, and label your prepared specimen in the space provided on the following page.

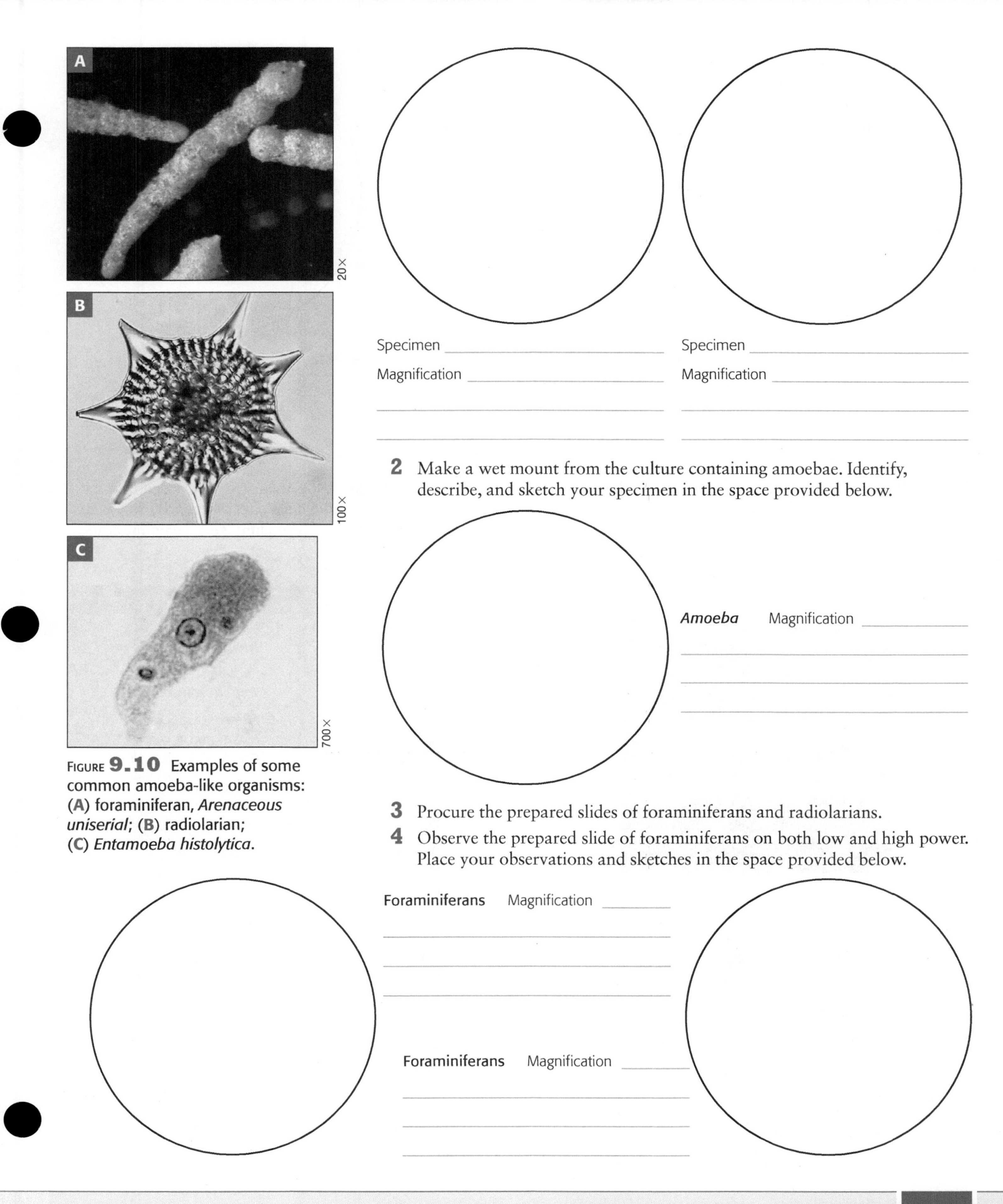


FIGURE **9.10** Examples of some common amoeba-like organisms: (**A**) foraminiferan, *Arenaceous uniserial*; (**B**) radiolarian; (**C**) *Entamoeba histolytica*.

**2** Make a wet mount from the culture containing amoebae. Identify, describe, and sketch your specimen in the space provided below.

***Amoeba*** Magnification ____________

**3** Procure the prepared slides of foraminiferans and radiolarians.

**4** Observe the prepared slide of foraminiferans on both low and high power. Place your observations and sketches in the space provided below.

**Foraminiferans** Magnification ____________

**Foraminiferans** Magnification ____________

**5** Observe the prepared slide of radiolarians on both low and high power. Place your observations and sketches in the space provided.

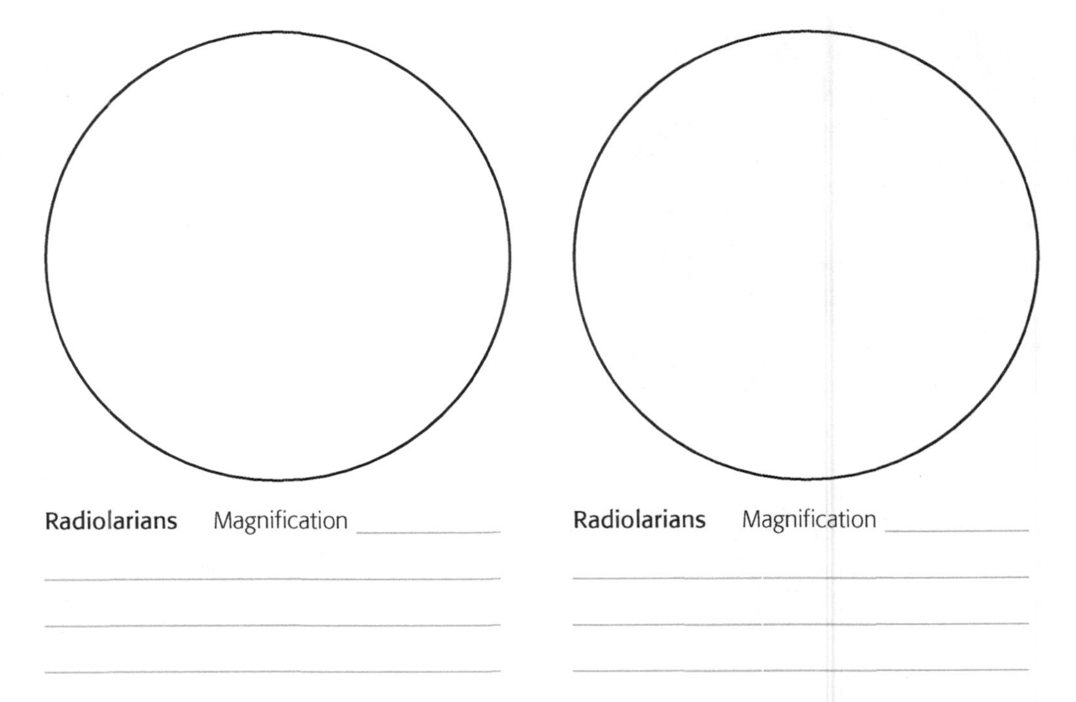


## Flagellated Protists

Perhaps the flagellated protists are the most confusing to taxonomists (Fig. 9.11). Classically, these are subdivided into the **phytomastigophorans** (photosynthetic flagellates) and the **zoomastigophorans** (animallike flagellates). Recently, they were placed into several supergroups, as indicated in Table 9.1. For convenience, the well-known **phylum Zoomastigophora** (**Sarcomastigophora**) is used here to describe the animallike protists that move by flagella.

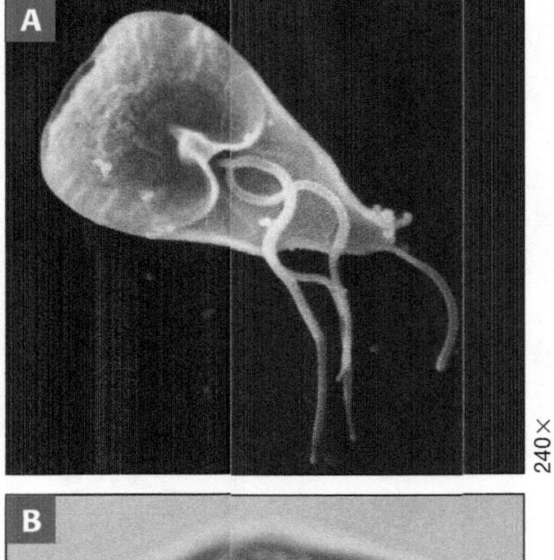


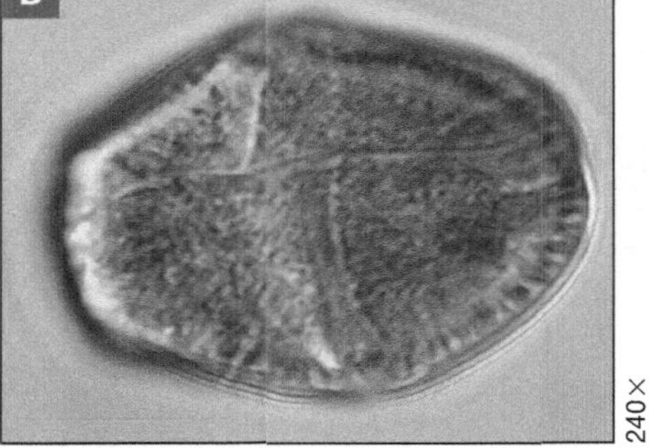


FIGURE **9.11** Flagellated protists: (**A**) *Giardia intestinalis (lamblia)*; (**B**) *Gymnodinium* sp.

TABLE **9.1** New View of the Protists

| Supergroup | Example | Representative Organism |
|---|---|---|
| Excavata | Diplomonadida<br>Parabasala | *Giardia intestinalis*<br>*Trichomonas vaginalis* |
| Euglenozoa (discicristates) | Euglenida<br>Kinetoplastea | *Euglena deses*<br>*Trypanosoma brucei* |
| Alveolata | Ciliophora<br>Apicomplexa<br>Dinoflagellata | *Paramecium caudatum*<br>*Plasmodium vivax*<br>*Gonyaulax catenella* |
| Amoebozoa | Amoebas<br>Cellular slime molds<br>Plasmodial slime molds | *Amoeba proteus*<br>*Dictyostelium discoideum*<br>*Physarum polycephalum* |
| Stramenopila (heterokonts) | Bacillariophyta<br>Oomycota<br>Chrysophyta<br>Phaeophyta | *Cytotella stelligera*<br>*Phytophthora infestans*<br>*Dinobryon sociale*<br>*Macrocystis pyrifera* |
| Rhizaria (cercozoa) | Radiolaria<br>Foraminifera<br>Chlorarachniophyta | *Lychnaspis miranda*<br>*Globigerina falconensis*<br>*Chlorarachnion reptans* |
| Archaeplastida | Rhodophyta<br>Chlorophyta<br>Glaucophyta<br>Vascular plants* | *Gelidium amansii*<br>*Spirogyra crassa*<br>*Cyanophora paradoxa*<br>*Acer rubrum* |
| Opisthokonta | Choanomonada<br>Fungi*<br>Animalia* | *Monosiga brevicollis*<br>*Amanita verna*<br>*Homo sapiens* |

* A separate kingdom

## Phylum Zoomastigophora (Sarcomastigophora)

The vast majority of zoomastigophorans are free-living, residing in freshwater, the marine environment, or in the soil. Several species are well-known parasites.

### Trypanosomes

The trypanosomes are protists responsible for a number of diseases in humans and other vertebrates. *Trypanosoma brucei* is responsible for causing African sleeping sickness. The vector for this disease is the tsetse fly. *Trypanosoma cruzi* is the infective agent that causes Chagas disease in South and Central America. The vector is the triatomine, or kissing bug. Chagas disease kills approximately 50,000 people annually.

Trypanosomes are endoparasites transmitted by arthropod vectors, and they live in the blood plasma of their host, so they are called **hemoflagellates.** Trypanosomes are characterized anatomically as having an elongated shape with a flagellum supported by microtubules originating in the posterior region. Trypanosomes possess a **kinetoplast** (a mass of mitochondrial DNA close to the nucleus) near the kinetosome. A kinetosome is a self-duplicating structure at the base of the flagellum. It is derived from a large mitochondrion and forms the basis for the potentially new kingdom Kinetoplastea. In addition, an undulating membrane is prominent in trypanosomes.

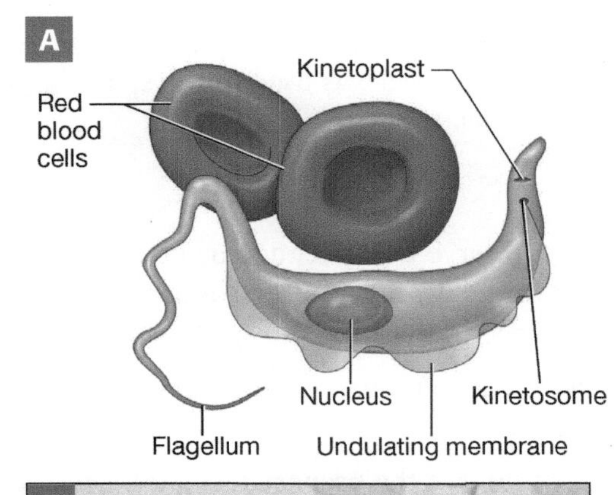


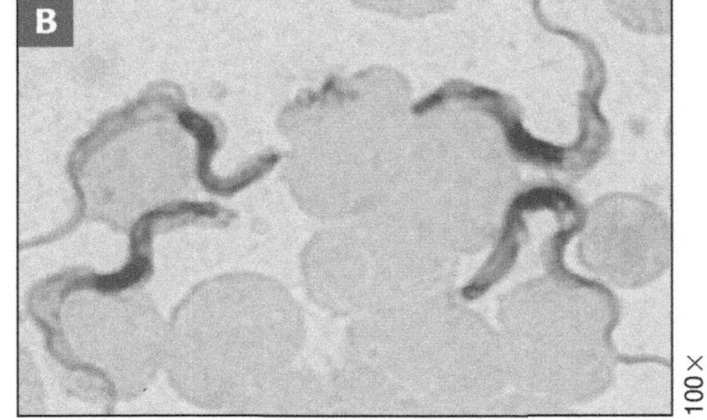


FIGURE 9.12 (A) Basic trypanosome anatomy; (B) trypanosomes in the blood plasma of a host.

### Other Zoomastigophorans

*Giardia intestinalis* (*lamblia*) causes giardiasis, an intestinal infection that causes intense diarrhea and dehydration. Giardiasis is a common waterborne infection worldwide. Many *Giardia* infections result from swimming in or drinking from contaminated waterways or even swimming pools. *Giardia* also is interesting in that it has modified mitochondria called mitosomes that do not generate ATP directly. In modern schemes, *Giardia* occupies its own kingdom known as **Diplomonadida.**

Another unusual flagellate is *Trichomonas vaginalis*, a sexually transmitted parasite that causes urogenital infections. *Trichomonas* possesses modified mitochondria, termed *hydrogenosomes*, and is placed in its own kingdom called **Parabasala.**

Many biologists also consider the **dinoflagellates** to be zoomastigophorans, but several schemes place them in the plantlike protist phylum **Pyrrophyta.** The dinoflagellates are primarily marine, but a few freshwater species exist. These protists possess a faceted, hard, cellulose case and a single, long flagellum. Although some are capable of photosynthesis, most are heterotrophs. Some, such as *Gymnodinium breve* and *Gonyaulax catenella*, are responsible for the devastating "red tide" in warm marine waters. *Gonyaulax* also can cause paralytic shellfish poisoning that can be lethal to humans.

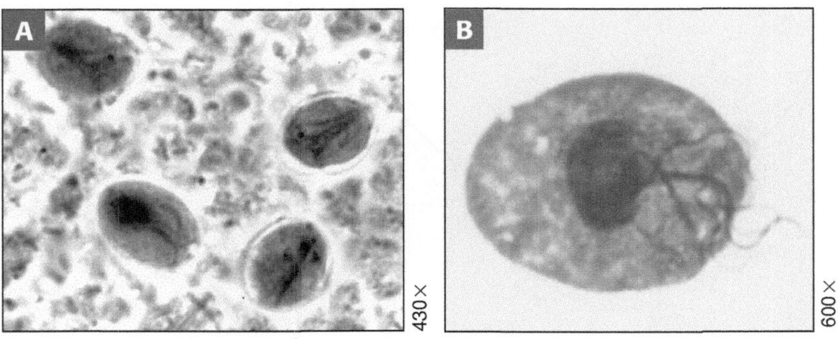


FIGURE 9.13 Two other types of flagellates: (A) *Giardia* (*intestinalis*) *lamblia*; (B) *Trichomonas vaginalis uniserial.*

In this procedure, you will observe representative flagellates from prepared slides and record your observations. First, you will view trypanosomes (Fig. 9.12). You will also observe *Giardia lamblia (intestinalis), Trichonympha* spp., and *Gonyaulax catenella* (Fig. 9.13). Finally, you will examine *Trichonympha* spp., which is essential to termites. This flagellate, living in a symbiotic relationship with termites, helps them digest cellulose. If termites are available, trichonymphs can be easily observed in their gut (Fig. 9.14).

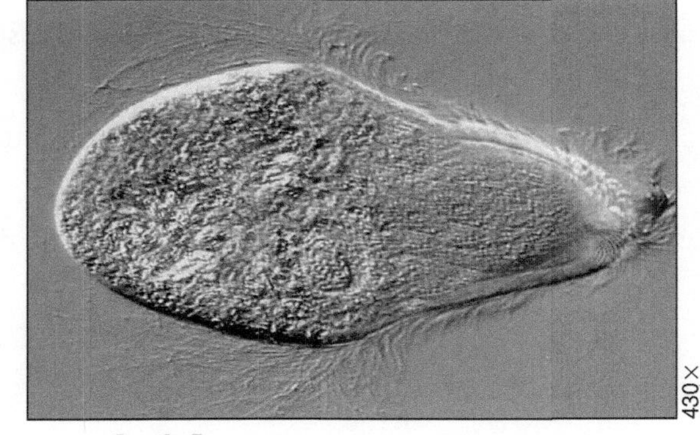


FIGURE 9.14 *Trichonympha* sp., a flagellate.

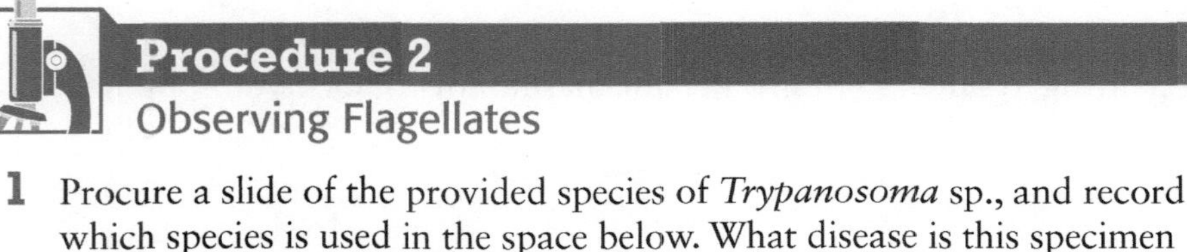

## Procedure 2

### Observing Flagellates

**Materials**

- ❑ Compound microscope
- ❑ Prepared slide of *Trypanosoma* sp.
- ❑ Prepared slide of *Gonyaulax catenella*
- ❑ Prepared slide of *Giardia (intestinalis) lamblia*
- ❑ Termites
- ❑ Microscope slides and coverslips
- ❑ Colored pencils

**1** Procure a slide of the provided species of *Trypanosoma* sp., and record which species is used in the space below. What disease is this specimen associated with?

______________________________

**2** Observe the specimen using low and high power. Locate the flagellum, undulating membrane, and nucleus. Draw and label your specimen in the space provided below.

**3** Procure prepared slides of *Gonyaulax catenella* and *Giardia* (*intestinalis*) *lamblia*.

**4** Observe the prepared slide of *Gonyaulax catenella* on both low and high power. Place your observations and sketches in the space provided below.

______________________________

______________________________

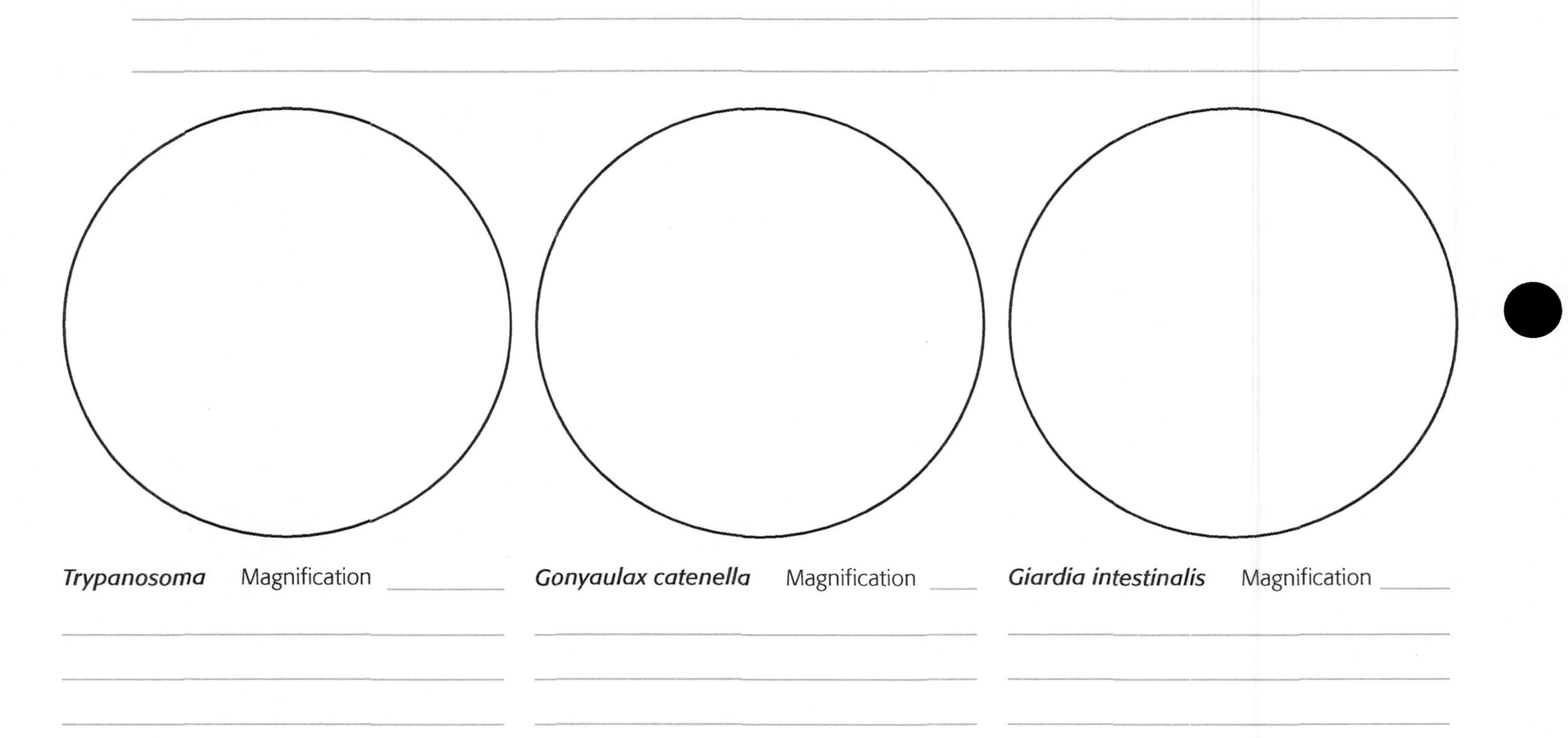

*Trypanosoma* Magnification ________ *Gonyaulax catenella* Magnification ________ *Giardia intestinalis* Magnification ________

**5** Observe the prepared slide of *Giardia* (*intestinalis*) *lamblia* on both low and high power. Place your observations and sketches in the space provided above.

**6** Write a brief paragraph describing the natural history of each organism you observed.

______________________________

______________________________

______________________________

______________________________

______________________________

7 Procure the equipment and a termite from the instructor. Place the termite on the center of a slide, and quickly place a coverslip over the termite. Press down on the coverslip, squashing the termite. Search through the remains of the termite. In the former gut region, you should see these unusual protists. Sketch *Trichonympha* spp., and place your observations in the space provided here.

8 Clean your laboratory area, properly dispose of the slide and coverslip, and wash your hands thoroughly.

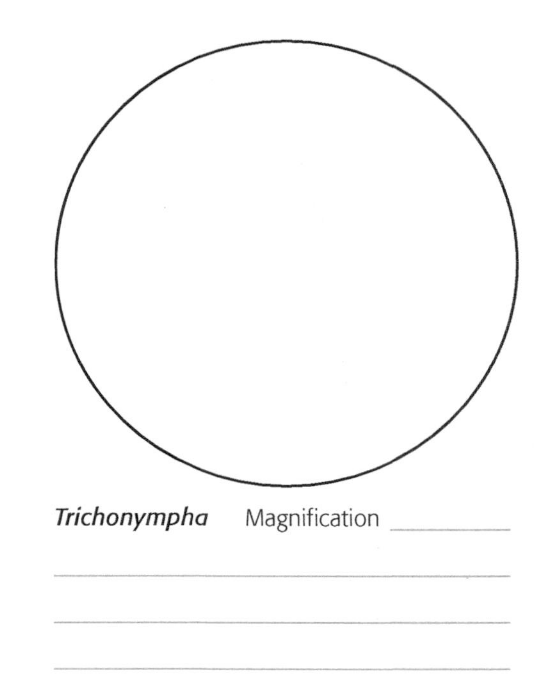

*Trichonympha* Magnification ________

________

________

________

## Ciliated Protists

### Phylum Ciliophora

Phylum **Ciliophora** consists of approximately 8,000 species of freshwater and marine organisms called **ciliates**. Their name is derived from the hairlike projections called **cilia** that cover their body surface or oral region. These structures are used in locomotion and feeding. The majority of ciliates are free-living, but there are several species of sessile (attached to a substrate), colonial, and parasitic forms.

The ciliates are considered the most structurally complex animallike protists. The outer covering, or pellicle, of ciliates is usually extremely tough. Most ciliates obtain their food using a **cytostome** (mouth) and an **oral groove.** The oral groove opens into a short canal, the cytopharynx, and ultimately into a region where food vacuoles are formed. The cytoproct, posterior to the oral groove, is responsible for elimination of fecal material. Contractile vacuoles are involved in osmoregulation. **Radiating canals** collect fluids and empty into the contractile vacuole.

Ciliates are multinucleate—possessing at least one **macronucleus** and one **micronucleus.** Macronuclei perform the basic nuclear duties, and the micronuclei are exchanged during conjugation, a form of sexual reproduction. Some ciliates possess trichocysts and toxicysts. Trichocysts lie beneath the pellicle and can be discharged like a harpoon for defense. Toxicysts release a poison that can immobilize prey.

The classic example of a free-living ciliate is *Paramecium caudatum* (Fig. 9.15) a common inhabitant of freshwater environments, such as ditches and ponds. Paramecia are more abundant in environments containing aquatic plants and decaying organic matter. The paramecium is slipper-shaped, transparent, and colorless. Paramecia appear highly active under the microscope, reaching speeds of 1 to 3 mm per second. These organisms have a pointed posterior end and a rather blunt anterior end. The oral groove is distinct, running backward to the ventral side. Paramecia can be observed using their cilia to channel food into the oral groove. The living pellicle is distinct and is covered with cilia.

In paramecia, the cilia near the oral groove create currents that bring water, and potentially food, into the oral groove. Trichocysts are used to gather food and for defense in paramecia. Contractile vacuoles in the paramecium act as an osmoregulatory apparatus, or bilge pump. Living in a freshwater (hypotonic) environment, the organism has to remove excess water that diffuses through the cell membrane.

*Stentor* spp., a sessile, vase-shaped ciliate, and *Vorticella* are also found in the freshwater environment. Freshwater fishes may suffer from "ich" caused by the ciliate *Ichthyophthirius multifiliis* (Fig. 9.15).

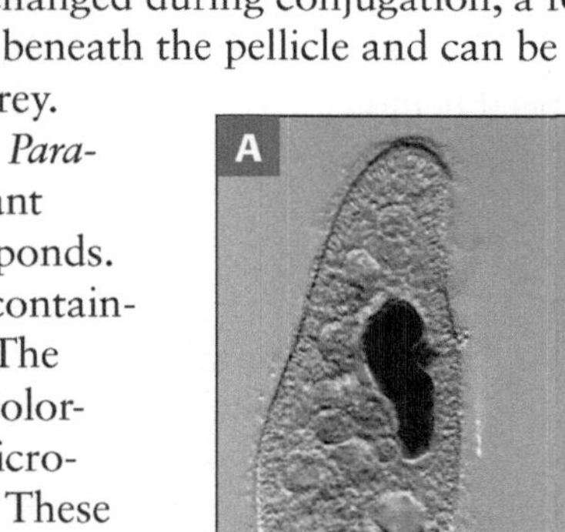


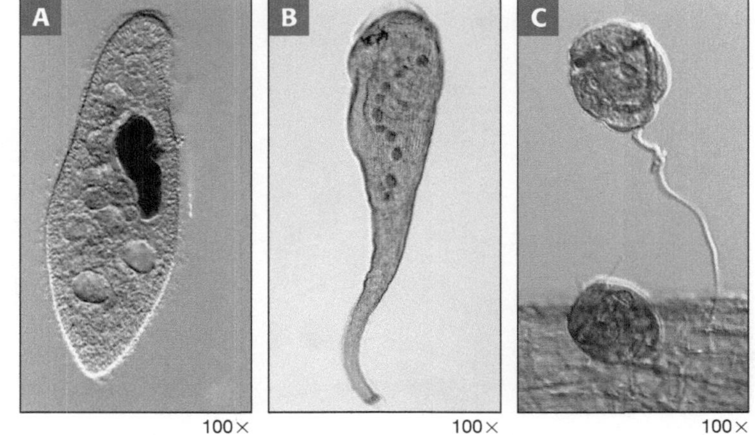

FIGURE **9.15** Examples from phylum Ciliophora: (**A**) *Paramecium caudatum*; (**B**) *Stentor* sp.; (**C**) *Vorticella* sp., common freshwater ciliates; (**D**) *Ichthyophthirius multifiliis*, known as "ich," seen here as small white specks on a freshwater fish.

## Procedure 3
### Observing Ciliates

In this procedure, you will examine a common ciliate protozoan known as *Paramecium caudatum* (Fig. 9.16).

**Materials**

- ❑ Compound microscope
- ❑ Prepared microscope slides of paramecia and conjugating paramecia
- ❑ Eyedropper
- ❑ Culture of *Paramecium* sp.
- ❑ Blank slides and coverslips
- ❑ Pipettes
- ❑ Vinegar
- ❑ India ink
- ❑ Protoslo®
- ❑ Probe
- ❑ Yeast solution stained with methylene blue
- ❑ Prepared slide of *Paramecium* conjugation
- ❑ Colored pencils

**1** Procure the equipment and material listed. Examine the prepared slide of *Paramecium caudatum* under both low and high power. Be able to identify the following structures: pellicle, cilia, trichocyst, oral groove, macronucleus, micronucleus, and contractile vacuole. Draw and label your specimen in the space provided.

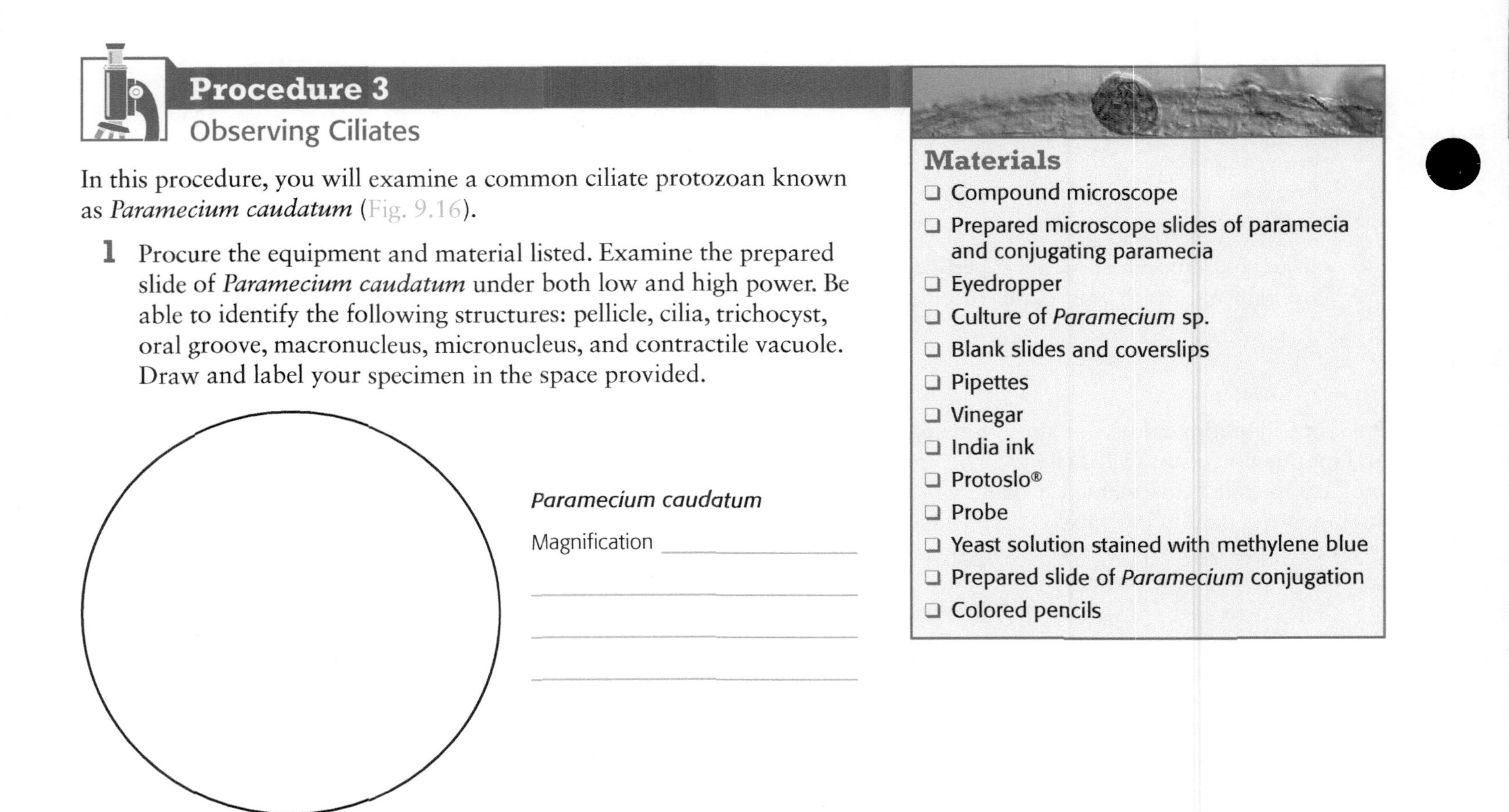

*Paramecium caudatum*

Magnification ______________

______________

______________

______________

**2** Prepare a wet mount of paramecia, and describe their movement, the action of the contractile vacuole, and the organism in general. Paramecia are "speed demons," so Protoslo® or methyl-cellulose can be used to slow them down significantly.

______________________________________________

______________________________________________

______________________________________________

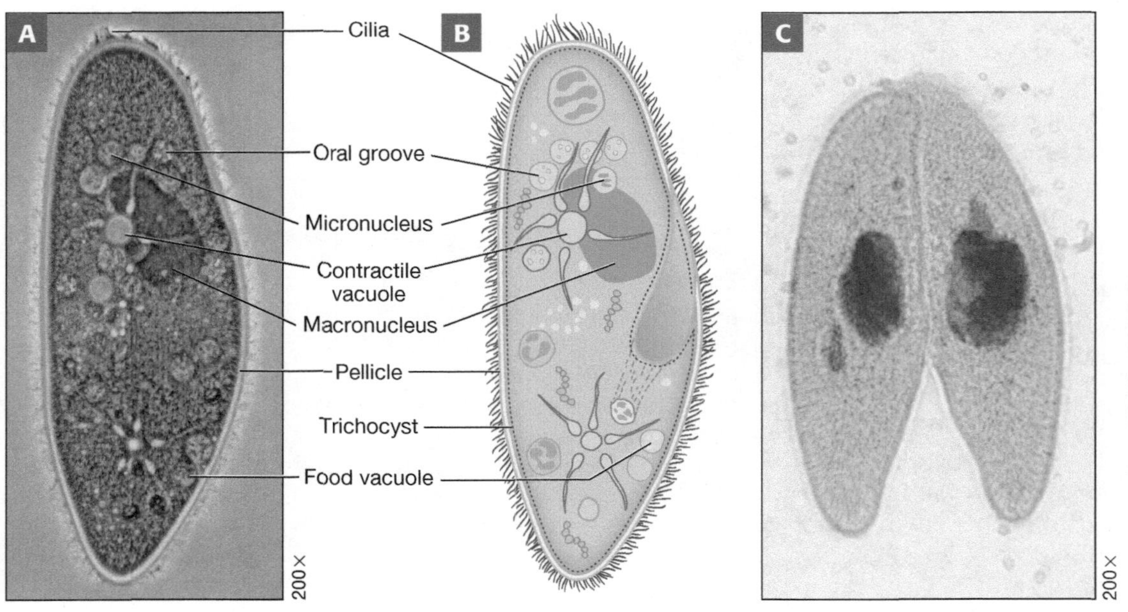


FIGURE **9.16** The paramecium a closer look: (**A**) *Paramecium caudatum*, (**B**) illustration of *Paramecium caudatum*; (**C**) *Paramecium* sp. in conjugation.

Name Date Section

## MYTHBUSTING

### What a Mess!

*Debunk each of the following misconceptions by providing a scientific explanation. Write your answers on a separate piece of paper.*

1. Red tide will not harm people.
2. I would not worry about these so-called emerging viruses and other diseases!
3. Since cats can harbor *Toxoplasma*, if a woman is pregnant, she should give her cat away.
4. We should not worry about giardiasis in developed countries.
5. All species of mosquitos can transmit malaria.

**1** What is the future of the kingdom Protista?

**2** Draw and label the anatomy of:

*Paramecium* spp. Magnification ______

*Euglena* Magnification ______

**3** How have the protists traditionally been classified? Provide an example of each type.

**4** Describe three parasitic protists.

**5** What is a pseudopod?

**6** Why should pregnant women be cautious with cats?

**7** The next time that you see a dead fish in an aquarium covered with white, cotton-looking stuff, what is that strange organism?

# There's a Fungus among Us
## Understanding Fungi

*Mushrooms are miniature pharmaceutical factories, and of the thousands of mushroom species in nature, our ancestors and modern scientists have identified several dozen that have a unique combination of talents that improve our health.*

—**Paul Stamets (1955–present)**

## OBJECTIVES

*At the completion of this chapter, the student will be able to:*

1. Explain the ecological role and characteristics of kingdom Fungi.
2. Describe and identify the basic anatomical features of kingdom Fungi.
3. Discuss the taxonomical organization of kingdom Fungi.
4. Describe the characteristics of and biology of phyla Chytridiomycota, Zygomycota, Glomeromycota, Ascomycota, and Basidiomycota.
5. Describe and identify select examples of phylum Zygomycota, Ascomycota, and Basidiomycota.
6. Trace the life cycle of *Rhizopus stolonifera*, a typical ascomycete, and *Agaris bisporus*, a typical basidiomycete.
7. Label the parts of a typical mushroom.
8. Describe the biology of lichens and their three basic forms.

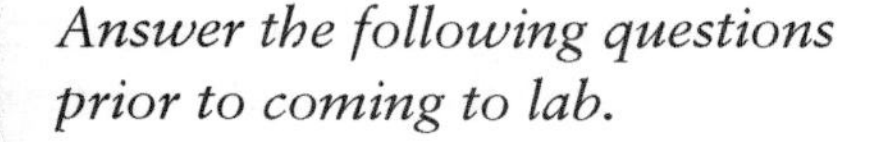


**1** Why do some fungi smell like rotten flesh?

______________________________________________

______________________________________________

**2** What is so special about a honey mushroom (*Armillaria ostoyae*) found in Malheur National Forest in eastern Oregon?

______________________________________________

______________________________________________

**3** What is the "going price" for truffles?

______________________________________________

______________________________________________

**4** What is the nutritional value of mushrooms?

______________________________________________

______________________________________________

**5** What is the relationship between bagpipes and *Aspergillus*?

______________________________________________

______________________________________________

Certainly, there are numerous species of fungi among us (Fig. 9.17)! Unfortunately, the term *fungus* evokes images of molds, mildew, rotting organic matter, spoiled food, and various maladies of plants, animals, and humans. Fungi do not limit their enzymatic attack to living or dead things. Species of fungi attack plastic, leather, paint, petroleum products, film, and even the multicoating of optical equipment such as cameras. Millions of dollars are spent yearly trying to control fungal diseases in plants, including Dutch elm disease, wheat rust, and corn smut; and in humans, diseases such as ringworm (red and itchy skin), coccidiomycosis (fever, rash, headache, joint pain, skin lesions, chronic pneumonia) and aspergillosis (allergic and lung infections). Some species of fungi are capable of producing powerful toxins, carcinogens, and hallucinogens.

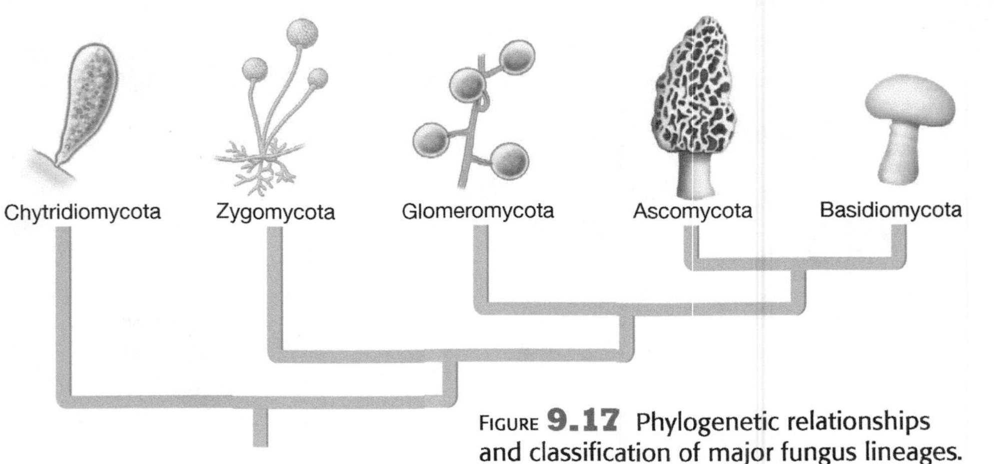


FIGURE **9.17** Phylogenetic relationships and classification of major fungus lineages.

However, fungi play a vital role in ecosystems and are economically essential (Fig. 9.18). Most fungi are **saprobes** that break down organic matter, but there are parasitic and mutualistic fungi as well. Without certain species of fungi serving as decomposers, ecosystems would collapse. These decomposers break down dead organisms, leaves, feces, and organic matter and recycle their chemical components back into the environment. In addition, many species of plants depend upon mutualistic fungi (mycorrhizae) to help their roots absorb minerals and water from the soil. Animals and humans eat many species of fungi. Truffles and some species of morels and mushrooms are delicacies. Fungi also play a vital role in the bread, cheese, beer, and wine industries. Several species of fungi are used in the production of antibiotics, including penicillin and cyclosporine, and other beneficial medicines.

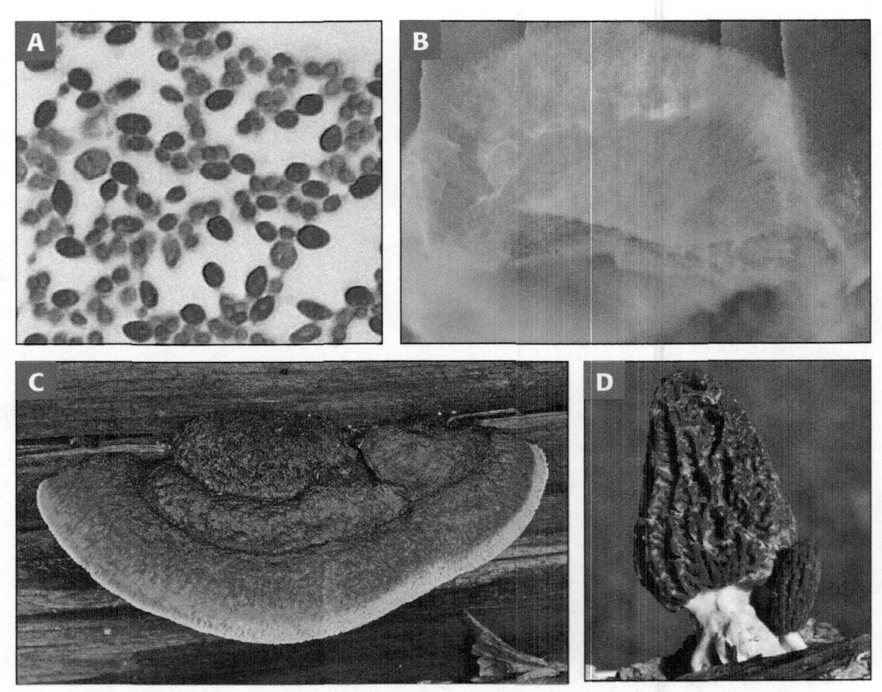


FIGURE **9.18** Examples of fungi: (A) yeast; (B) bread mold; (C) mushroom; (D) morel.

## Anatomical Structure

Fungi are filamentous, spore-producing, heterotropic unicellular or multicellular eukaryotes. Fungi release digestive enzymes onto a food source, partially dissolving the source to make the essential nutrients available. Multicellular fungi are composed of numerous small filaments, known as **hyphae**, grouped together into a mass called a **mycelium**. For example, hyphae make up the body of a mushroom. Hyphae and mycelia can grow rapidly. Under ideal conditions, a single fungus might produce a kilometer of hyphal growth in one day. That is why a mushroom can appear overnight.

In the majority of fungi, the hyphae are divided into compartments by **septa** (cross walls) that allow for some structures, such as ribosomes and mitochondria, to pass. Fungi known as coenocytic fungi lack septa and appear as a cytoplasmic mass with perhaps thousands of nuclei. Various modifications of hyphae can be found throughout the fungi. One modified hyphae, a haustoria, penetrates the tissues of a host. Other modified hyphae, called **rhizoids**, anchor fungi to a substrate. Some soil fungi have snare, or loop, hyphae to trap unsuspecting nematodes (roundworms) for future consumption. Fungi

that have cross walls in their hyphae are connected to adjacent hyphae by tiny pores in the cross wall; in contrast, septa that separate reproductive cells have no pores. Many fungal species possess a cell wall composed of the polysaccharide chitin, and they store their energy in the form of glycogen. Fungi do not contain chlorophyll.

Most fungi are capable of undergoing asexual and sexual reproduction. Asexually, fungi can reproduce by budding, fragmentation, and spore formation. Spores can develop directly without uniting with another spore. Sexually, fungi produce gametes in specialized areas of the hyphae called **gametangia**. The gametes may be released to fuse into spores elsewhere, or the gametangia themselves may fuse. In the hyphae of some fungi, **dikaryons** (Greek, *di* = two, *karyon* = nucleus) form as the result of unspecialized hyphae fusing. In this case, their two nuclei remain distinct for a portion of the life cycle. When the two nuclei finally fuse, the zygote undergoes meiosis prior to spore formation. Upon germination, the spores form haploid hyphae. A large mushroom can produce billions of spores. When a spore lands on a suitable substrate, it germinates and grows. That is why bread mold can appear mysteriously on a slice of bread you thought was pristine.

## Classification

For many years, the fungi were classified as imperfect plants. Today, it is clear the fungi are not degenerate plants but, rather, unique eukaryotes deserving of their own unique kingdom. Modern classification schemes place fungi closer to the animal kingdom than to the plant kingdom. Mycologists (specialists in fungi) recognize nearly 100,000 species of fungi and predict this number could increase to 2 million species. Representative fungi include mushrooms, puffballs, bread mold, morels, truffles, smuts, rusts, blight, mildew, and yeasts.

Members of kingdom Fungi are classified into five distinct phyla based on anatomy, types of hyphae, means of reproduction, and molecular biology:

1. Chytridiomycota is the most ancient group of fungi. Most chytrids are either aquatic decomposers feeding on dead plant or animal material in a pond or parasites living on water molds, insects, frogs, or snakes. Several chytrids can cause a fatal disease in frogs and are responsible for the decline of frog populations in several regions on Earth.
2. Zygomycota includes bread molds and *Pilobolus* spp., the "hat-throwing" fungus.
3. Glomeromycota comprises fungi that enter into a symbiotic relationship with the roots of plants, forming mycorrhizae. Perhaps as many as 95% of the land plants are associated with mycorrhizae. In the relationship the fungi receive carbohydrates, sucrose, and glucose from the plant. The fungi also help the plants take in phosphate and other minerals. The mycorrhizae were important to the evolutionary success of terrestrial plants in conquering the land.
4. Ascomycota, the largest phylum of fungi, includes organisms such as truffles, morels, and yeast.
5. Basidiomycota includes mushrooms, shelf fungi, and puffballs.

In this chapter, we will be reviewing zygomycetes, ascomycetes, and basidiomycetes. Phyla Chytridiomycota and Glomeromycota will not be covered in this chapter.

## Phylum Zygomycota

Phylum Zygomycota includes 1,000 species of primarily terrestrial fungi known as the conjugating fungi. **Zygomycetes** commonly occur in soil, decaying organic matter (a white filamentous mass on decaying fruit), and feces. The hyphae of zygomycetes lack septa and are called coenocytic. A representative example of phylum Zygomycota is *Rhizopus stolonifer*, the common black bread mold. Three types of hyphae are found in *Rhizopus*:

1. rhizoids (anchoring hyphae that penetrate the bread and have digestive enzymes)
2. stolons (horizontal surface hyphae)
3. sporangiophores (reproductive hyphae)

Reproduction in *Rhizopus* can occur asexually or sexually. Asexually, when a **spore** (sporangiospore) lands on a suitable substrate, such as a slice of bread, it germinates and forms hyphae that soon form a mycelium. After the mycelium develops, it produces sporangiophores that rise above the surface and contain spore-containing **sporangia**. The sporangia release their spores and seek another supportive substrate.

Sexually, *Rhizopus* reproduces by conjugation. *Rhizopus* produces two different hyphae (strains) that develop swollen **progametangia** on the ends facing each other. Eventually they touch, and a cross wall forms behind each tip. Next, a thick-walled **zygosporangium** forms, replacing the progametangia. The zygosporangium cracks open, forming sporangiophores and their associated sporangia. Meiosis occurs in the sporangia, producing **meiospores** then released to seek another substrate.

Several other notable species of zygomycetes are medically and ecologically significant. *Rhizopus nigricans*, which also can grow on bread, and *Mucor* spp., which can grow on stored foods and seeds and are commonly found in house dust, causing fungal sinusitis and allergies. Other species of *Rhizopus* and *Mucor* can cause serious and sometimes deadly infections (mucormycosis) that affect the skin, digestive tract, facial region, lungs, and brain. *Pilobolus*, discussed in the Nature's Cannon sidebar, is another zygomycete. Once considered protists, the **microsporidia** are placed with zygomycetes in some classification schemes, whereas in others, they constitute a separate phylum of kingdom Fungi (Microsporidia). The microsporidia are capable of infecting several species of animals including humans. They are particularly devastating to individuals with compromised immune systems.

Did you know . . . ?

### Nature's Cannon: Hitching a Ride

Grazing animals such as cattle rarely graze near feces (dung pat). The zone of ungrazed grass around a dung pat is called the "ring of repugnance." Infective stages of several endoparasites are faced with the task of having to travel from the dung pat beyond the ring of repugnance to ungrazed grass.

The roundworm (nematode) *Dictyocaulus viviparus*, called the "cattle lungworm," has solved this problem in a unique way. The larvae of the cattle lungworm migrate up the sporangiophore of the saprobic fungus *Pilobolus* sp. and accumulate on the sporangium.

*Pilobolus* is known as the shotgun fungus, or the hat-throwing fungus, or the cannon fungus because it has explosive sporangia that can shoot spores beyond the ring of repugnance up to 2.5 meters toward light. The lungworm larvae hitch a ride on the spores and land beyond the ring of repugnance. When ingested by a cow, the larvae penetrate the wall of the cow's intestine and are carried by the lymphatic and circulatory systems to the lungs, where the adult worms develop.

## Procedure 1

### Macroanatomy of *Rhizopus stolonifer*

**Materials**

- ❑ Dissecting microscope or hand lens
- ❑ Scalpel
- ❑ Paper towels
- ❑ Water
- ❑ Bread contaminated with *Rhizopus stolonifer*
- ❑ Colored pencils

**1** Procure the equipment and the bread mold from your instructor.

**2** Using a dissecting microscope or hand lens, observe your specimen. Attempt to find all of the anatomical structures shown in Figure 9.19. Sketch, label, and record your observations in the space provided below.

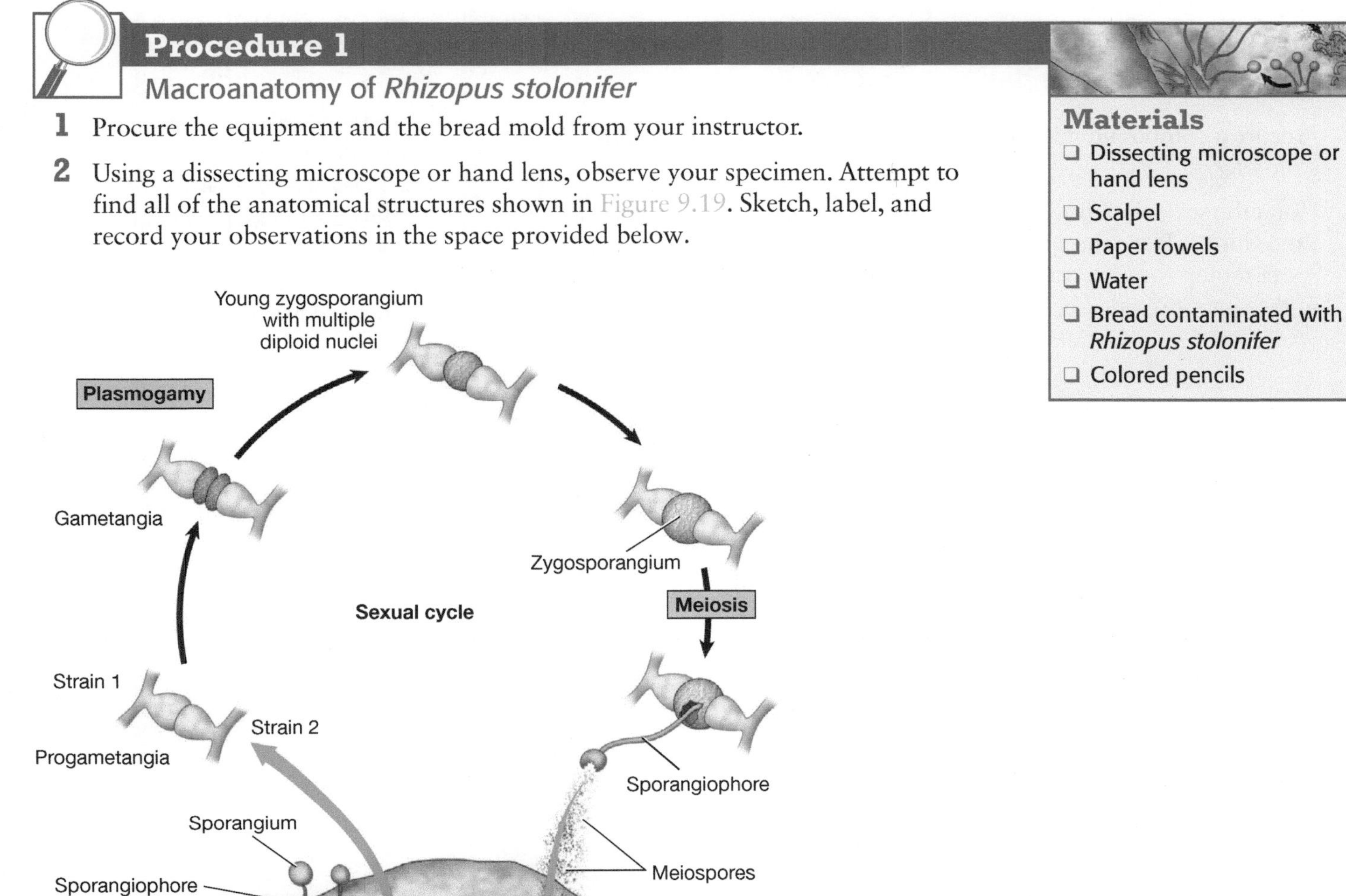


**Figure 9.19** Life cycle of *Rhizopus stolonifer*.

*Rhizopus stolonifer* ______________________

______________________

______________________

______________________

## Procedure 2
### Microanatomy of *Rhizopus stolonifer*

**Materials**
- ❑ Compound microscope
- ❑ Slides and coverslips
- ❑ Prepared slides of *Rhizopus stolonifer*
- ❑ Scalpel
- ❑ Colored pencils

**1** Procure a compound microscope and prepared slides of *Rhizopus stolonifer*.

**2** Using the scalpel, remove a sporangium from the sporangiophore in Procedure 1. Place the sporangium on a microscope slide, and prepare a wet mount for observation. Sketch, label, and record your observations in the space provided below.

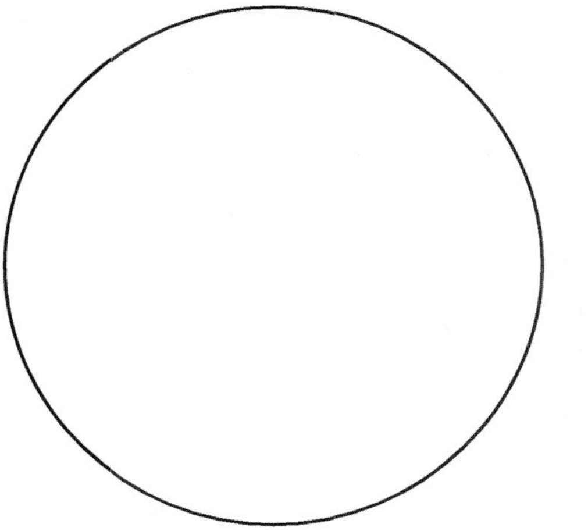

*Rhizopus stolonifer*

Magnification ________________

**3** Examine the prepared slides and compare them with Figures 9.20 and 9.21. Place your observations and labeled sketches in the space provided on the following page.

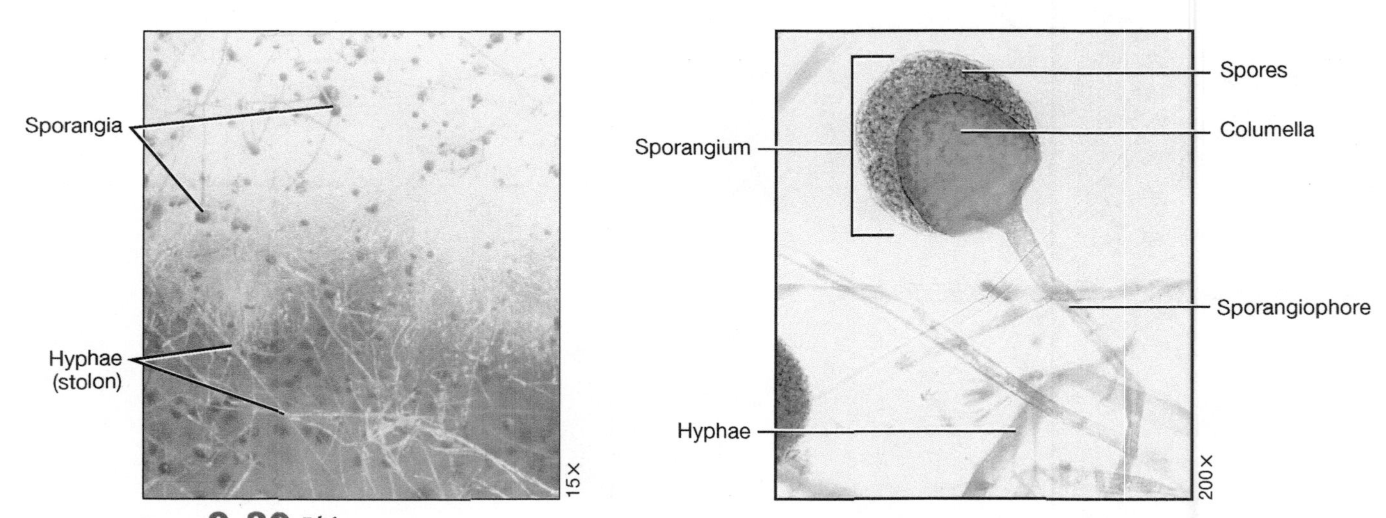


FIGURE **9.20** *Rhizopus* sp.

FIGURE **9.21** Whole mount of the bread mold, *Rhizopus* sp.

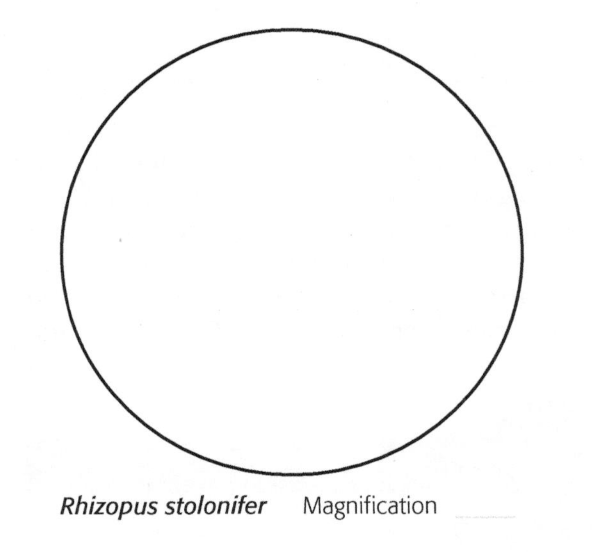

*Rhizopus stolonifer* Magnification ______

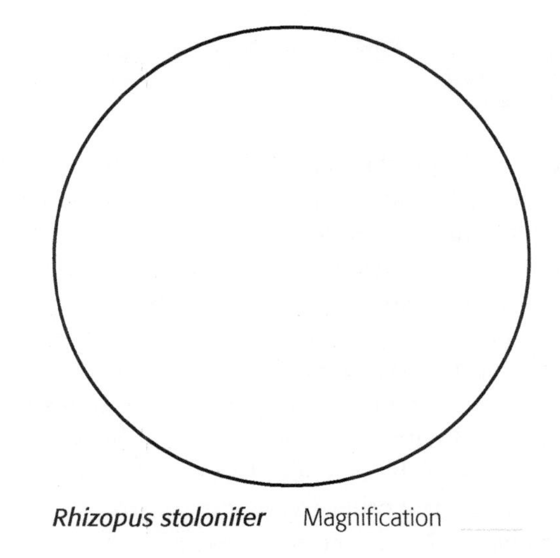

*Rhizopus stolonifer* Magnification ______

## Check Your Understanding

**1.1** How do zygomycetes get their name?

**1.2** What is a sporangium?

**1.3** How does a slice of bread become contaminated with bread mold?

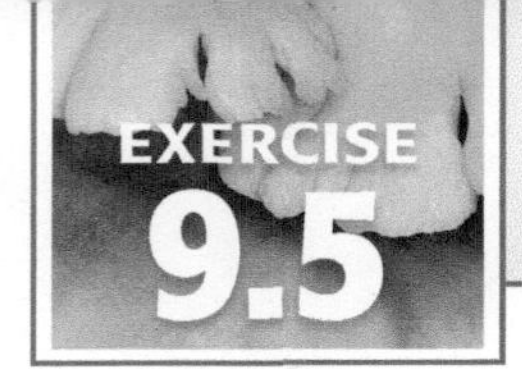

# EXERCISE 9.5 Phylum Ascomycota

Most described fungal species (65,000) belong to phylum Ascomycota, the sac fungi (Fig. 9.22). Sac fungi live in a variety of marine, freshwater, and terrestrial habitats. These organisms range from unicellular to elaborate multicellular forms. **Ascomycetes** are responsible for several serious plant diseases such as powdery mildew, chestnut blight, and Dutch elm disease. Coccidiomycosis is a fungal disease in humans caused by an ascomycete, resulting in rash and potentially deadly pulmonary disease. *Cryptococcus neoformans* is an ascomycete associated with pigeon excretions and is potentially deadly to animals and humans. It is associated with meningitis and meningioencephalitis. *Histoplasma capsulatum*, another ascomycete, is responsible for the potentially lethal lung disease known as histoplasmosis. *Claviceps* spp. are associated with St. Anthony's fire in the Middle Ages, the madness of the witches of Salem, the hallucinogenic drug LSD, and other cases of ergotism throughout history. Many species of ascomycetes are found in a symbiotic relationship with algae, forming **lichens**.

The yeast *Saccharomyces cerevisiae*, an ascomycete, plays important roles in the brewing industry and in genetic research. Another yeast, *Candida albicans*, causes a variety of fungus infections, including oral thrush and vaginal infections. *Neurospora*, a type of bread mold, is also important in genetic studies. True morels are common woodland ascomycetes featuring a distinguishing convoluted cap. Several species of morels are prized for their flavor and consistency. Keep in

> **Did you know . . . ?**
>
> **Ergot . . . Madness!**
>
> One of the most interesting ascomycetes, *Claviceps purpurea*, or ergot, grows on rye and similar plants. *Claviceps* is responsible for ergotism in humans and other animals that consume infected food. In the Middle Ages, the dreaded St. Anthony's fire was caused by ergot. Ever since the Middle Ages, ergot has been used to induce abortions and to stop maternal bleeding following childbirth. Ergot produces a chemical used to synthesize lysergic acid diethylamide, or LSD. Perhaps the "possessed" and "mad" people in Salem were merely "tripping out" as the result of ergotism.

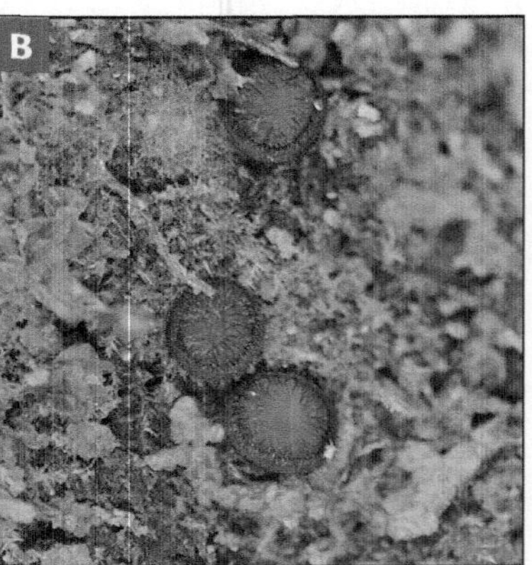


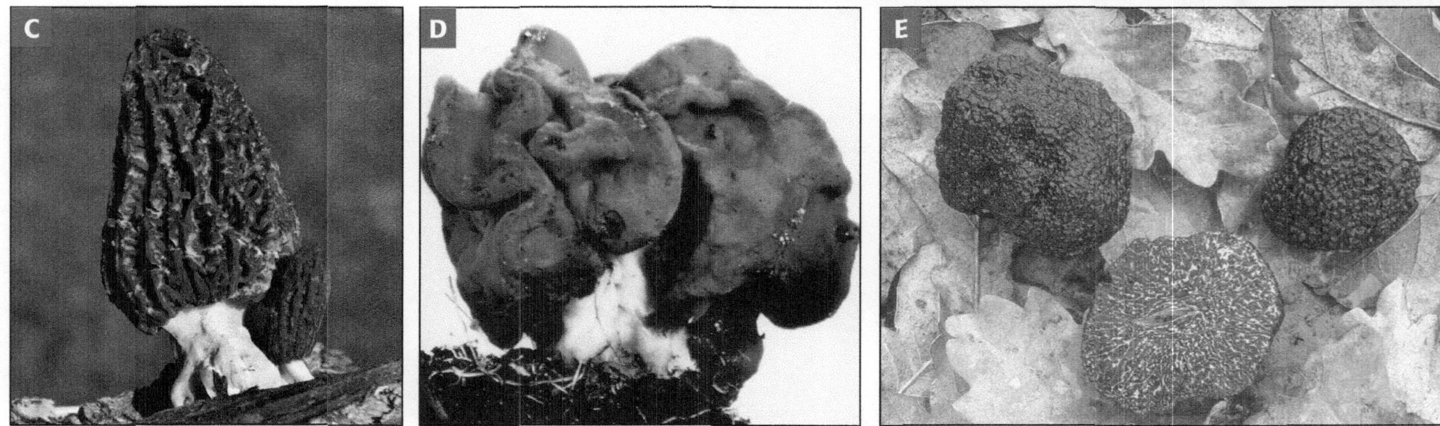


FIGURE **9.22** Fruiting bodies (ascocarps or ascoma) of common ascomycetes: (**A**) *Peziza repanda* is a common woodland cup fungus. (**B**) *Scutellinia scutellata* is commonly called the eyelash cup fungus. (**C**) *Morchella esculenta* is a common edible morel. (**D**) *Helvella* is sometimes known as a saddle fungus because the fruiting body is thought by some to resemble a saddle. (**E**) Truffle.

mind that several species of morels are poisonous and only an expert should harvest morels for human consumption. Although plain to the sight, truffles have an exquisite flavor and rival the cost of some precious metals per gram.

*Aspergillus* is a genus of green mold that can cause deadly respiratory infections. Some species of *Aspergillus* are used in producing soy sauce, toothpaste, chewing gum, inks (especially black), and photograph-developing solutions. The *Aspergillus flavus* species that may grow on improperly stored grain produces a potent carcinogenic substance that can cause liver cancer. *Stachybotrys chartarum* is a black mold responsible for "sick-building" syndrome, in which exposure to the spores can cause chronic sickness such as headaches, eye irritation, lung disease, rash, memory loss, and fever. This mold presented a major problem in New Orleans, Louisiana, and the Mississippi gulf coast (Gulfport and Biloxi) after Hurricane Katrina.

*Penicillium chrysogenum*, formerly called *Penicillium notatum,* is the source of penicillin, the antibiotic discovered fortuitously by Alexander Fleming in 1928. Several species of *Penicillium* are used in the production of gourmet cheeses, such as Brie, Camembert, Gorgonzola, and Roquefort. What do you think the "blue stuff" in blue cheese is (Fig 9.23)? But less-friendly species of ascomycetes, including *Trichophyton* sp., cause athlete's foot, ringworm (Fig. 9.24), jock itch, and small nonpigmented splotches of skin called tinea versicolor.

Ascomycetes get their name from the **ascus**, a large, saclike cell responsible for producing reproductive **ascospores.** The hyphae in ascomycetes are septate, but the cross walls are not complete. Fruiting bodies in ascomycetes are well developed and are called **ascocarps.** Sexual reproduction in the ascomycetes starts when hyphae with one nucleus of opposite mating strains come into contact (Fig. 9.25). Each female gametangium, called an **ascogonium**, forms a **trichogyne** that grows toward the male gametangium, called the **antheridium.** After the trichogyne touches the antheridium, nuclei migrate from the antheridium to the female ascogonium. The ascogonium forms dikaryotic **ascogenous hyphae.** These hyphae form a crozier, or hook, and the nuclei fuse, forming a diploid nucleus. The nucleus undergoes meiosis, producing eight ascospores. Eventually, the ascospores forming in the ascocup are released. The asexual spores form singularly or in chains from **conidiophores** and are called **conidia.**

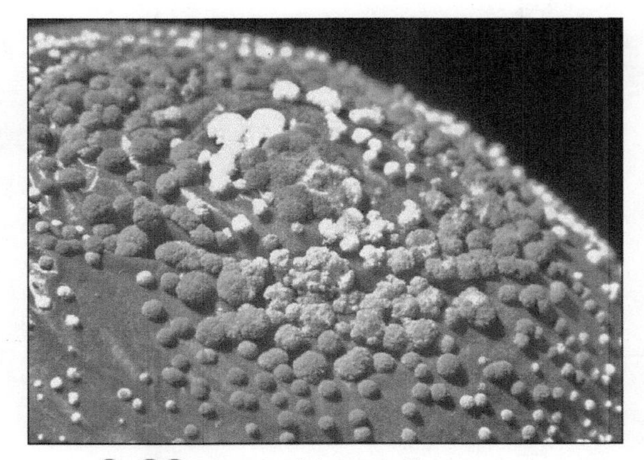

FIGURE **9.23** Blue mold, *Penicillium expansum*, growing on a rotten pear.

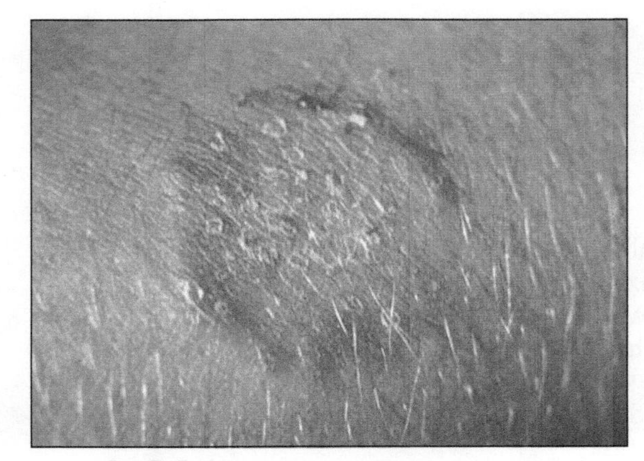

FIGURE **9.24** Ringworm, *Tinea corporis.*

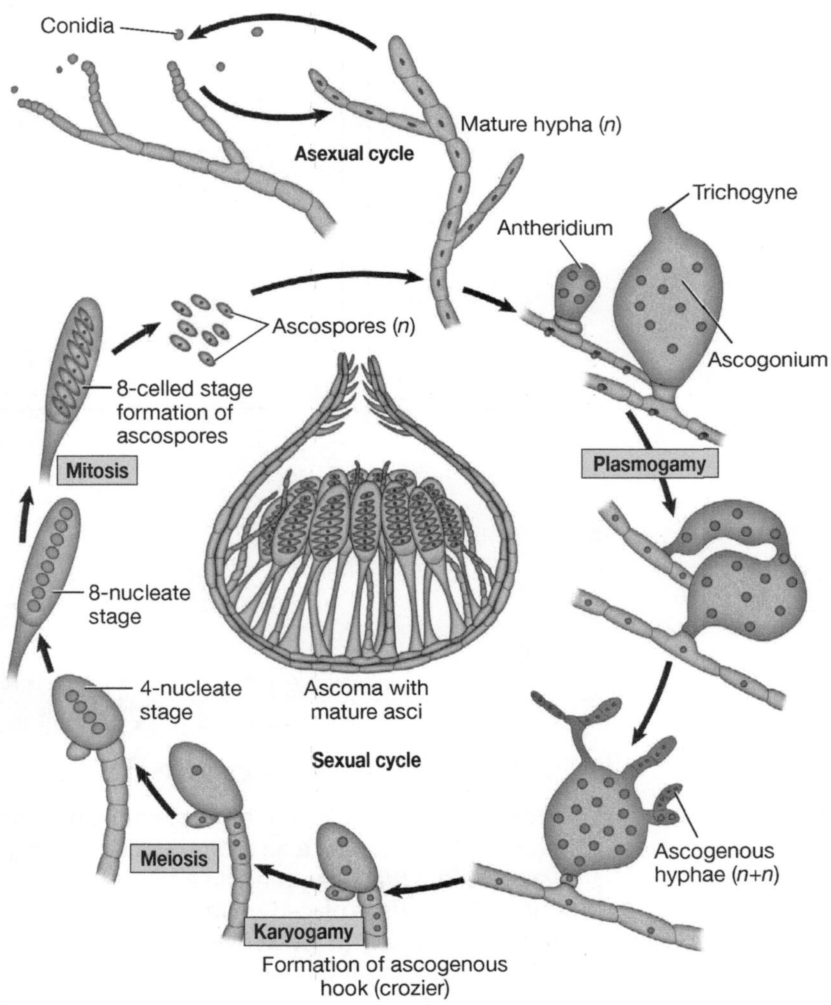


FIGURE **9.25** Life cycle of an ascomycete.

## Procedure 1
### Macroanatomy of Ascomycetes

**Materials**
- ❑ Dissecting microscope or hand lens
- ❑ Dissecting tray
- ❑ Scalpel
- ❑ Paper towels
- ❑ Water
- ❑ Several examples of true and false morels and plants infected with ascomycetes provided by the instructor
- ❑ Colored pencils

1. Procure the equipment and sample specimens.
2. Using a hand lens or dissecting microscope, observe your specimens. Record your observations and sketches in the space provided below.

Specimen ____________________

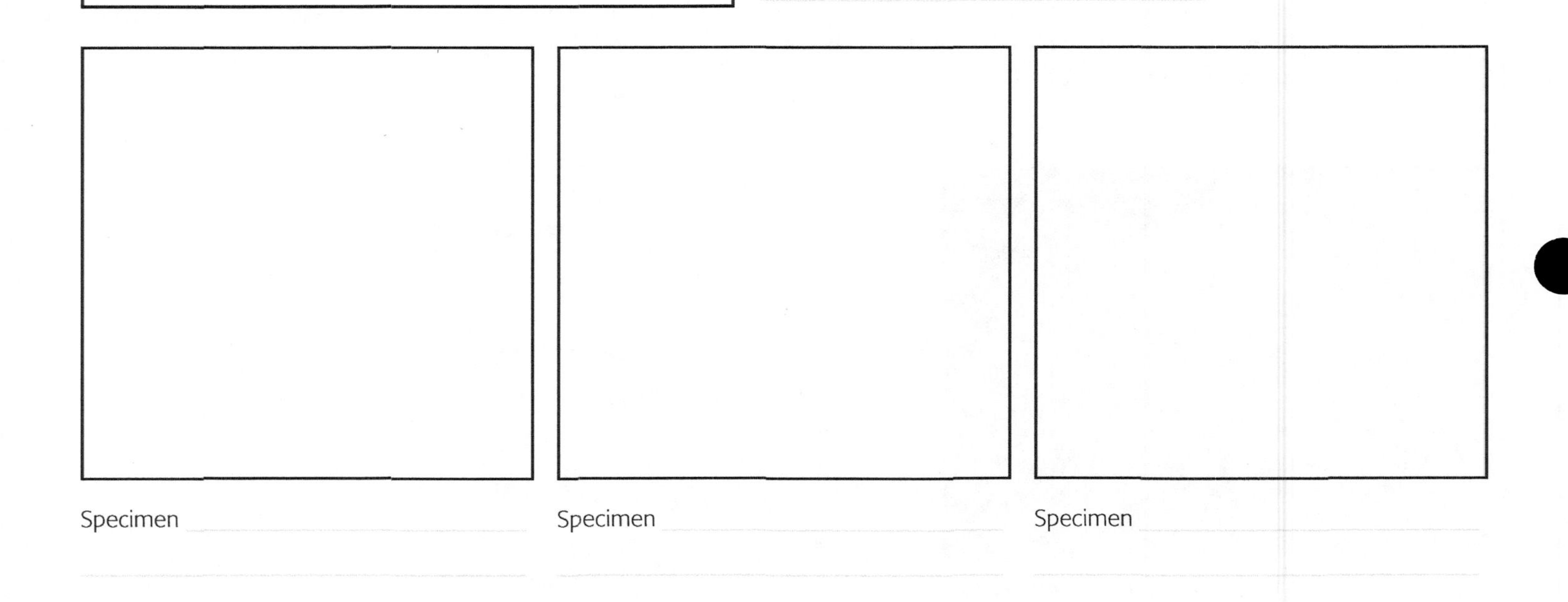

Specimen ____________________

Specimen ____________________

Specimen ____________________

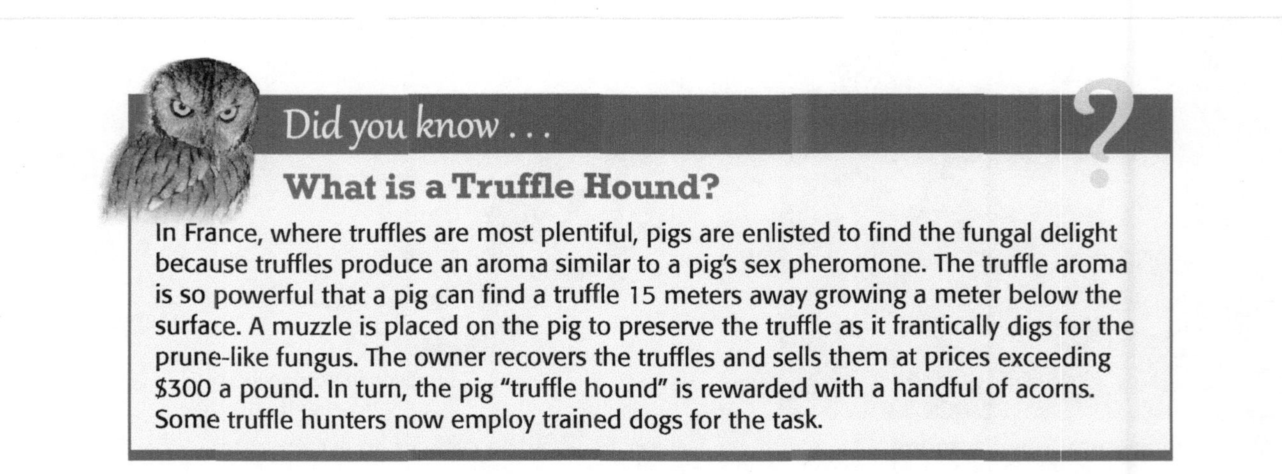

Did you know . . .

### What is a Truffle Hound?

In France, where truffles are most plentiful, pigs are enlisted to find the fungal delight because truffles produce an aroma similar to a pig's sex pheromone. The truffle aroma is so powerful that a pig can find a truffle 15 meters away growing a meter below the surface. A muzzle is placed on the pig to preserve the truffle as it frantically digs for the prune-like fungus. The owner recovers the truffles and sells them at prices exceeding $300 a pound. In turn, the pig "truffle hound" is rewarded with a handful of acorns. Some truffle hunters now employ trained dogs for the task.

## Phylum Basidiomycota

The best-known phylum of fungi is Basidiomycota, with more than 30,000 known species. Example **basidiomycetes** include mushrooms, puffballs, jelly fungi, earth stars, chanterelles, stinkhorns, shelf fungi, rusts, and smuts (Fig. 9.26). Basidiomycetes are called "club fungi" because they produce spores, **basidiospores**, in a club-shaped structure, the **basidium**. Most basidiomycetes are saprobes living on dead or dying plants.

Mushrooms, the most obvious basidiomycetes, may be seen living singly, in groups, or in a circle called a **fairy ring**. Several species of mushrooms, including portabella (*Agaricus bisporus*), oyster mushroom (*Pleurotus ostreatus),* shiitake (*Lentinula edodes*), and chanterelles (*Cantharellus cibarius*), are edible and known for their delectable taste.

One has to be careful when collecting mushrooms for consumption because many poisonous mushrooms resemble edible species to the untrained eye. Some poisonous mushrooms, such as *Amanita* spp., are colorful and appealing, yet this species is termed the "death angel" or "death cap" because of its poison. Some mushrooms are extremely hallucinogenic or psychedelic, such as the "magic mushroom," *Psilocybe* spp.

**Did you know . . . ?**

**Mushroom or Toadstool?**

In some circles, the common terms *mushroom* and *toadstool* cause confusion. Most Americans call any club-shaped fungus a mushroom. British people generally designate the term *mushroom* for edible fungi, and the inedible mushrooms as toadstools. Some people reserve the term *toadstool* for toxic fungi.

FIGURE **9.26** Representative basidiomycetes: (A) death angel, *Amanita* sp.; (B) basidiomycete puffballs, *Calvatia* sp.; (C) jelly fungus, *Tremella* sp.; (D) earthstar, *Astraeus* sp.; (E) golden chanterelle, *Cantharellus* sp.; (F) orange stinkhorn, *Clathrus* sp.; (G) corn smut, *Ustilago maydis*; (H) shelf fungus from the family Polyporaceae; (I) cereal rust, *Puccinia graminis*.

Puffballs are other common basidiomycetes. They literally release their spores into the wind when they split. Shelf, or bracket, fungi resemble small shelves growing on the trunk of a tree. Jelly fungi usually are colorful and feel cold, rubbery, or gelatinous to the touch. Stinkhorns are diverse, from orange, fingerlike structures erupting from the soil to something resembling a whiffle ball. If this were a scratch-and-sniff manual, everyone would agree the slimy covering of these fungi smells like rotting flesh. Rusts such as cedar-apple rust and wheat rust (*Puccinia triticina*) are parasitic fungi devastating to wheat and rye crops. Smuts such as *Ustilago maydis* are parasitic fungi that attack corn, sugarcane, and other cereal crops, resulting in much devastation.

The basidiomycetes reproduce primarily through sexual reproduction. The life cycle of a mushroom is typical of most basidiomycetes (Fig. 9.27). When a spore lands on a suitable substrate, it germinates into a network of hyphae that form a mycelium beneath the surface. Haploid hyphae exist in several reproductive types. When two compatible types unite, they form a new dikaryotic mycelium. These mycelia can live for perhaps a hundred years and spread, forming the underground surface of a fairy ring. The mycelia eventually form a button that emerges from the soil. The button develops into a typical mushroom, sometimes called a **basidiocarp** or **basidioma.**

A typical mushroom is composed of a cup-shaped **volva** at the base, a stalk-like structure called a **stipe**, a ring around the upper end of the stipe called an **annulus**, and a cap, or **pileus.** Beneath the cap are slit-like structures called **gills**, or they may be pore-like structures. The gills are composed of individual basidia. In immature mushrooms, a veil may cover the developing gills. The basidia mature, and the two nuclei fuse, forming a diploid nucleus that undergoes meiosis. The resulting four basidiospores can be found on peg-like **sterigma.** A large mushroom can produce several million basidiophores in a few days. The spores are released, and the cycle begins again.

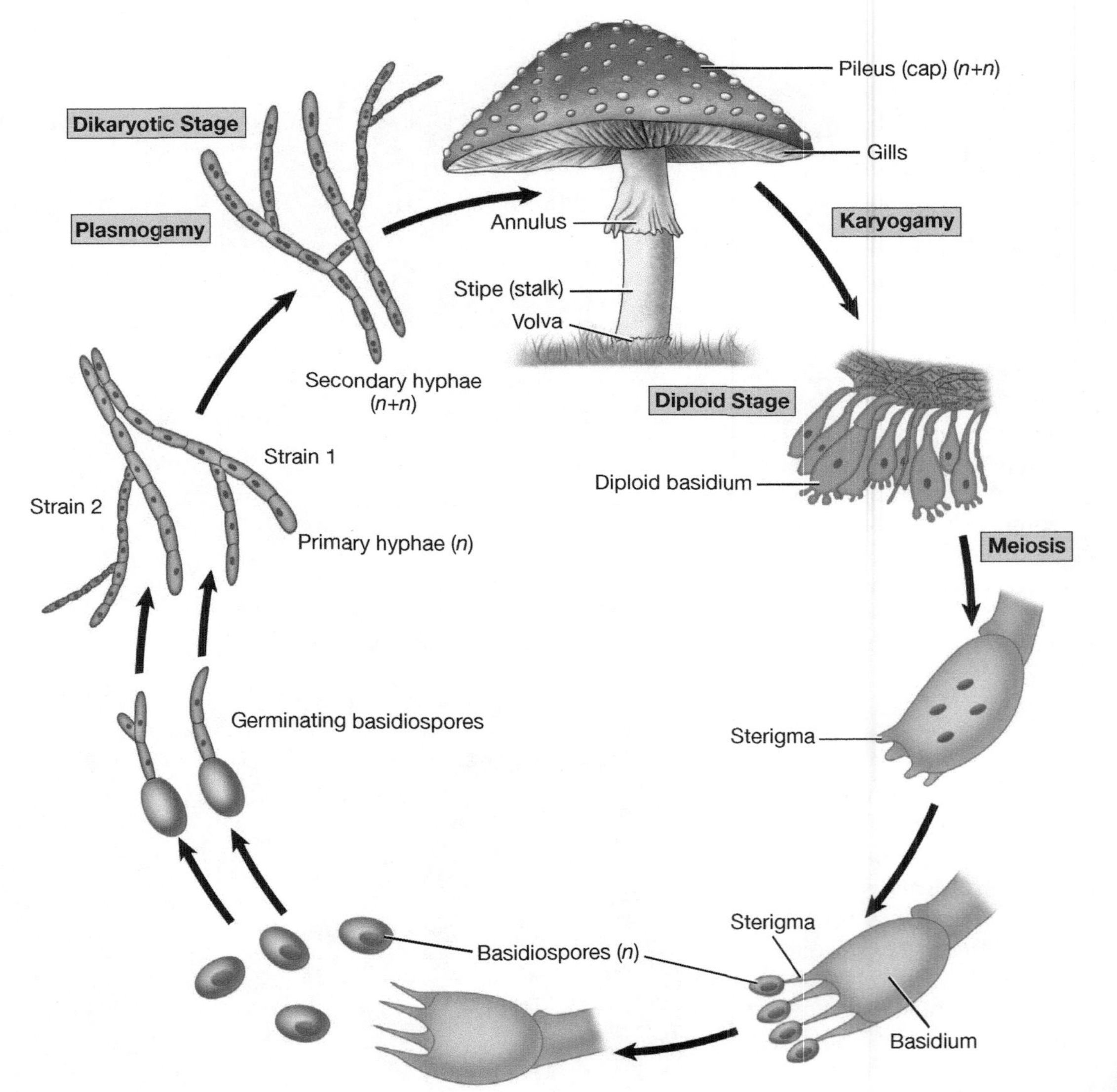


FIGURE **9.27** Life cycle of a "typical" basidiomycete (mushroom).

## Procedure 1

### Macroanatomy of Basidiomycetes

1 Procure the equipment and sample specimens.

2 Using a hand lens or dissecting microscope, observe your specimens (Fig. 9.28). Record your observations and sketches in the space provided below.

**Materials**

- ❑ Dissecting microscope or hand lens
- ❑ Scalpel
- ❑ Paper towels
- ❑ Water
- ❑ Several examples of basidiomycetes, such as mushrooms and puffballs
- ❑ Colored pencils

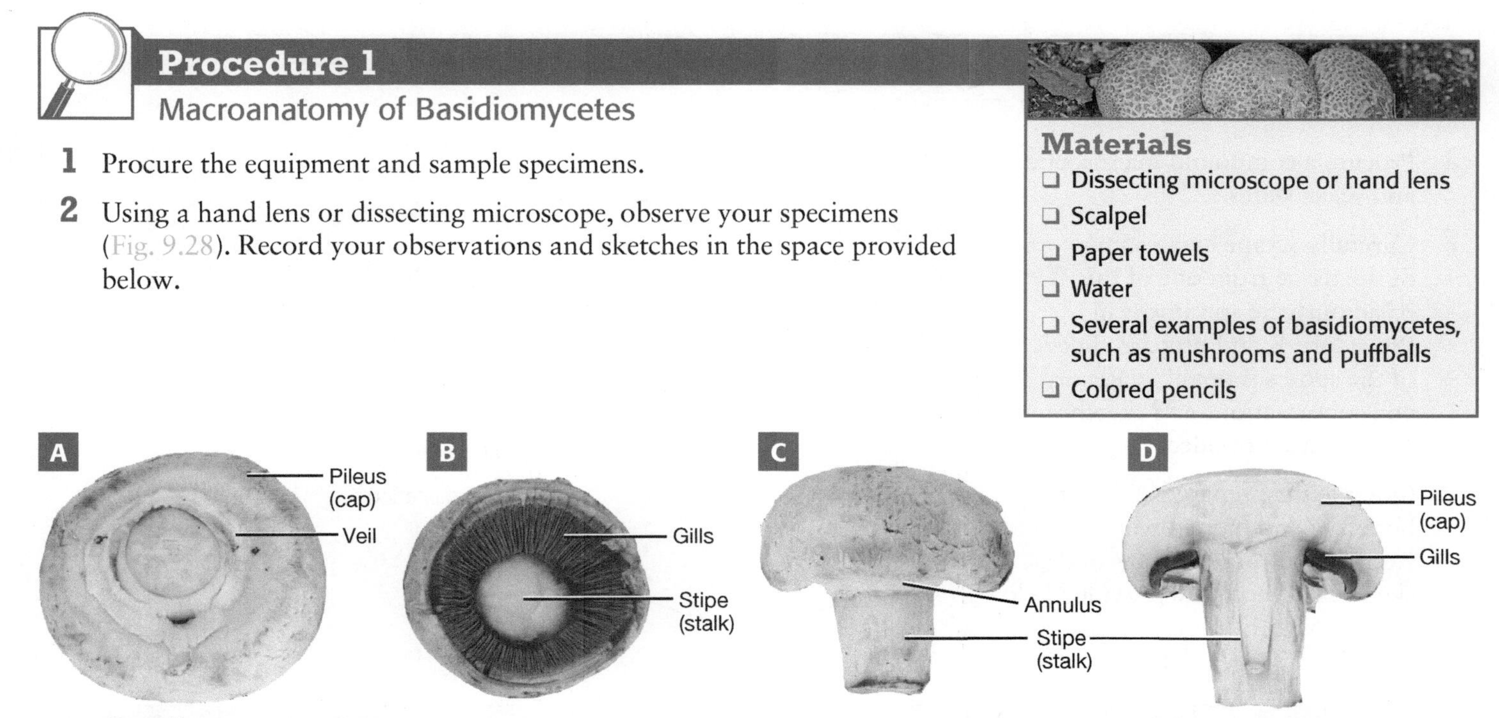


FIGURE **9.28** Structure of a mushroom: (**A**) inferior view with the annulus intact; (**B**) inferior view with the annulus removed to show the gills; (**C**) lateral view; (**D**) longitudinal section.

Specimen ______________________

Specimen ______________________

## Procedure 2
### Microanatomy of Basidiomycetes

**Materials**
- ❑ Compound microscope
- ❑ Slides and coverslips
- ❑ Prepared slides of select basidiomycetes, including a mushroom *Coprinus* sp.
- ❑ Colored pencils

1. Procure a compound microscope and select slides.
2. Carefully scrape away some of the tissue from one of the basidiomycete specimens in Procedure 1, and prepare slides of the spores if possible. Record your observations and sketches in the space provided.
3. Observe the prepared slides on both low and high power (Fig. 9.29). Record your observations and sketches in the space provided below and on page 129.

Specimen ______________

Magnification ______________

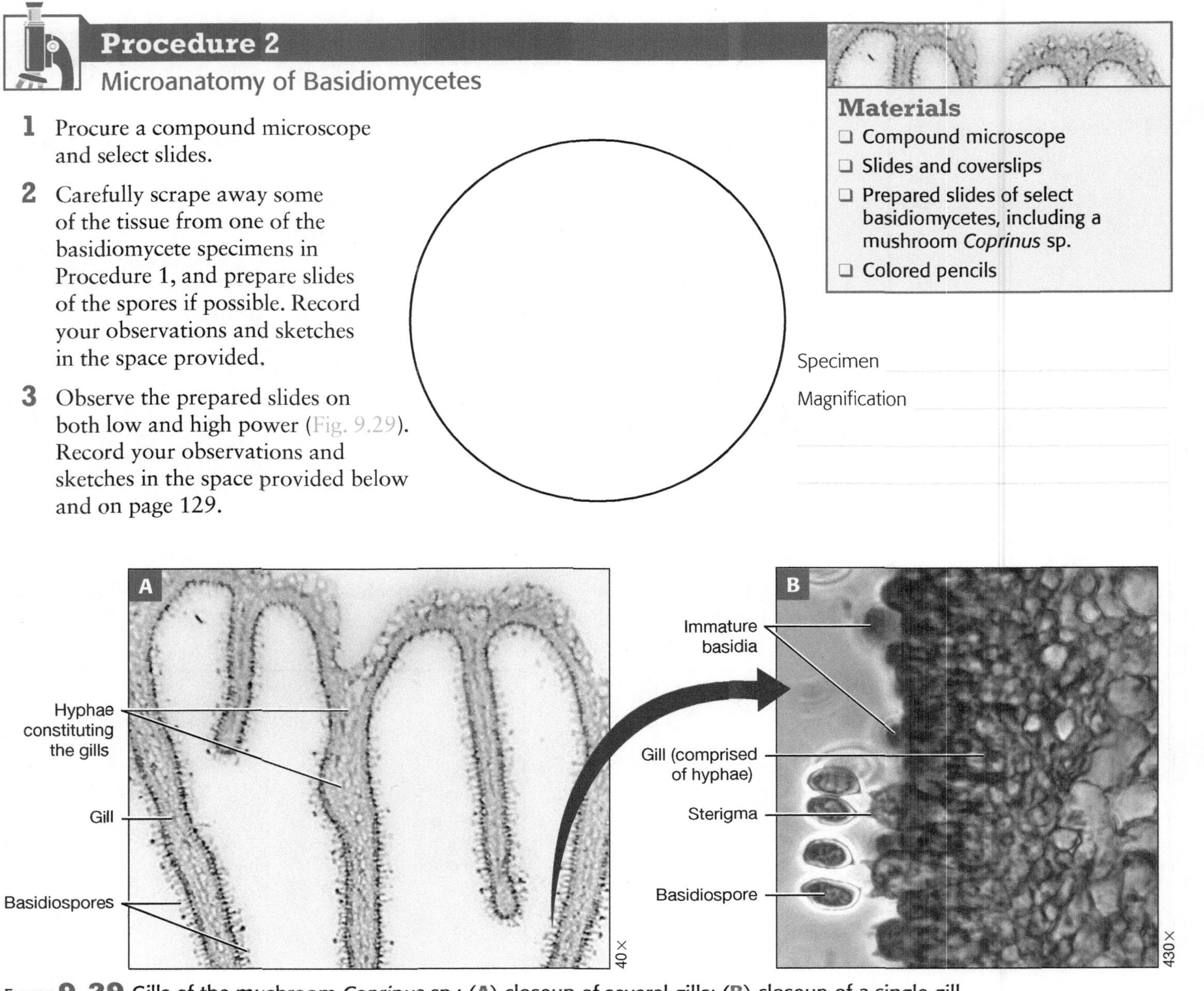

FIGURE **9.29** Gills of the mushroom *Coprinus* sp.: (**A**) closeup of several gills; (**B**) closeup of a single gill.

Specimen ______________

Magnification ______________

Specimen ______________

Magnification ______________

Specimen

Magnification

Specimen

Magnification

Specimen

Magnification

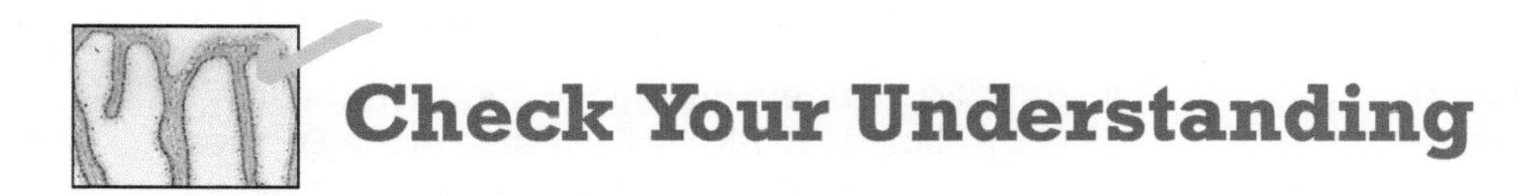

## Check Your Understanding

**3.1** Describe five types of basidiomycete.

**3.2** How did the basidiomycetes get their name?

**3.3** Describe two basidiomycete plant pathogens.

# Exercise 9.7 Lichens

Lichens are interesting symbionts consisting of a green algae or a cyanobacterium and, with the exception of a few species, ascomycetes. Algal cells or cyanobacteria are thought to provide food for both symbionts through photosynthesis, and the ascomycete retains water and minerals, anchors the organism, and protects the algae. Presently, nearly 20,000 species of lichen have been described. Scientific names are assigned to lichens like other species. Lichens typically reside on trunks and branches of trees, bare rocks, and human-made structures, such as walls and gravestones. They also can survive in extreme conditions, such as the tundra (e.g., reindeer moss) and hot deserts. Lichens have been used to make dyes, litmus paper, bandages, antibiotics, packing material, decorations, and perfume. In the environment, pioneer lichens help build soil, and they provide food and habitat for small animals. Some species of lichen serve as environmental indicators of air pollution. Recently, several species of nitrogen-fixing lichens have been described.

The body, or **thallus,** of a lichen is usually derived from an ascomycete surrounding algal cells and enclosing them within complex fungal tissues. The thallus ranges in size from less than 1 millimeter to more than 2 meters in diameter. Lichens are noted for their longevity, perhaps living 4,500 years. Lichens vary in color from dull gray to bright red, green, and orange. Lichens primarily reproduce asexually.

Three basic types of lichens exist in nature:

1. **Crustose** lichens form brightly colored patches, or crusts, on rock or tree bark, without evident lower surfaces.
2. **Foliose** lichens appear to have leaflike thalli that overlap, forming a scaly, lobed body. These lichens frequently are found on tree bark and on human-made structures.
3. **Fruticose** lichens may appear shrub-like or hanging mosslike on trees. Their thalli are either highly branched or cylindrical. Many people think lichens are parasites on trees but, with the exception of a few species, this is incorrect (Fig. 9.30).

> **Did you know . . . ?**
>
> **Lichens in Space: A Lesson in Being Tough**
>
> In 2005, lichens were exposed to the harsh conditions of space aboard the BIOPAN-5 section of the European Space Agency facility for 16 days. Fungal and algal cells of lichens were found to survive in space after full exposure to massive UV and cosmic radiation—conditions proven to be lethal to bacteria and other microorganisms. In addition, after being dehydrated as the result of a vacuum, the lichens recovered within 24 hours.

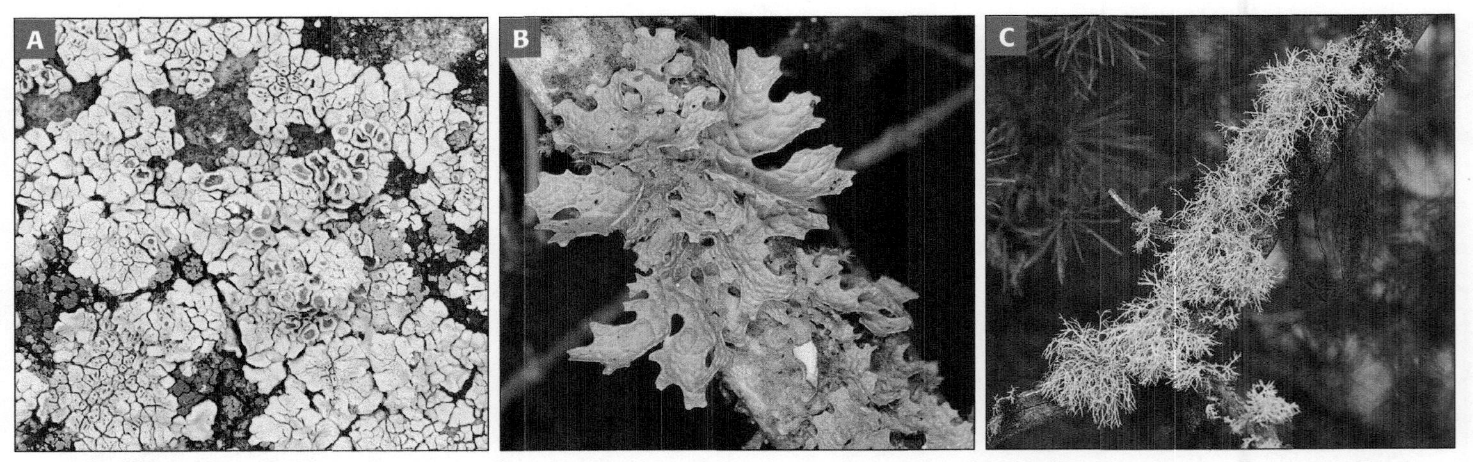


FIGURE **9.30** Lichens are often categorized informally by their form: (A) gold cobblestone, *Pleopsidium* sp., a crustose lichen; (B) cabbage lungwort, *Lobaria* sp., a foliose lichen; (C) wolf, *Letharia* sp., a fruticose lichen.

## Procedure 1

### Macroanatomy of Lichens

1. Procure the equipment and specimens.
2. Using a dissecting microscope or hand lens, observe, describe, and sketch the lichens in the space provided below.

**Materials**

- ❑ Dissecting microscope or hand lens
- ❑ Dissecting tray
- ❑ Several examples of crustose, fruticose, and foliose lichens
- ❑ Colored pencils

Specimen ______________________

Specimen ______________________

Specimen ______________________

## Check Your Understanding

**4.1** What is a lichen?

**4.2** Why are lichens important to the environment?

**4.3** Where can one find lichens?

Name ______ Date ______ Section ______

## MYTHBUSTING

**Break the Mold!**

*Debunk each of the following misconceptions by providing a scientific explanation. Write your answers on a separate piece of paper.*

1. If animals of the forest eat a particular mushroom, it is safe for me to eat.
2. I cannot get athlete's foot from a pair of used shoes that I buy online.
3. The presence of bread mold means that my house is filthy.
4. The false morel *Gyromitra,* or the beefsteak morel, contains jet fuel.
5. Ringworm is caused by parasitic worms.

**1** What is the significance of kingdom Fungi to the environment?

**2** Why are scientists concerned about certain species of chytrids?

**3** Name and describe five typical ascomycetes.

**4** Trace the generalized life cycle of an ascomycete.

**5** What are mycorrhizae and what is their evolutionary significance?

**6** What are five typical basidiomycetes?

**7** Trace the generalized life cycle of a basidiomycete.

**8** Draw and label a typical mushroom.

**9** What are three important ascomycetes?

**10** What is a lichen? Describe three types of lichens.

# Photosynthesis: Converting Light Energy into Chemical Energy

10

***Before coming to lab***

☐ Read lab

☐ Answer pre lab questions

## PRE LAB QUESTIONS

1. Differentiate between autotrophs and heterotrophs.

2. What is the summary equation for photosynthesis?

3. What is the plant that we will use to isolate pigments from?

4. What are the 4 pigments we will separate?

5. What procedure will we use to separate the pigments?

6. What is a spectrophotometer? What does it measure?

7. What are stomata?

## Purpose of the Lab

The purpose of this lab is to identify plant pigments by separating them using paper chromatography and to determine the absorption spectra of each of these pigments. Furthermore students will design and conduct their own experiments to study stomata in plant leaves.

**The sun** is the ultimate source of energy for the vast majority of organisms in the biosphere. Through a process called **photosynthesis,** plants and other **photoautotrophs** use the sun's energy to convert simple inorganic compounds into energy-rich organic molecules. **Autotrophs** (meaning "*self-feeders*" in Greek) are organisms that make their own food and sustain themselves without consuming organic molecules derived from other organisms. **Heterotrophs** (*hetero* meaning "other") consume other organisms as their source of energy. Some heterotrophs known as **herbivores** consume plants. **Carnivores** are heterotrophs that consume other animals. Heterotrophs known as **decomposers** obtain the energy-rich organic molecules needed for life from dead organisms.

**What is the summary equation for photosynthesis?**

______________________________

Photosynthesis is actually a series of chemical reactions that occur in **chloroplasts** in plant cells. During the **light reactions,** certain wavelengths of light energy are **absorbed** by **photosynthetic pigments** while other wavelengths are **reflected** or **transmitted.**

### Key Concept

**Is light absorbed, reflected or transmitted?**

The light that you can see passing through a leaf or a glass of water is light that is

______________________________.

The color that we see is the light that is ______________________________.

The light that provides the energy for photosynthesis is the light that is

______________________________.

## The Nature of Sunlight

The **electromagnetic energy** of sunlight travels through space as waves. The distance between the crest of two adjacent waves is called the **wavelength.** The following diagram shows the full range of electromagnetic energy from very short gamma rays to much longer radio waves.

Note that visible light is only a small fraction of this spectrum.

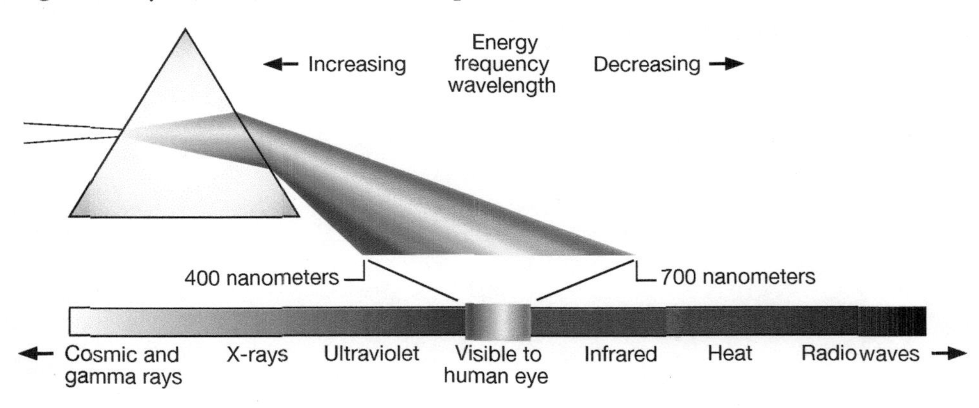

## The Relationship between Color and Wavelength

Check out the chart posted in the lab room that depicts the relationship of wavelength to color. On **your data table on** page 142 of this lab, record the name for **the color for each wavelength.** The names of the colors are standardized as **ROY G BIV** also known as red, orange, yellow, green, blue, indigo, and violet. The shortest wavelengths of visible light are associated with the color ______________________ while the longest wavelengths of visible light appear as the color ______________________.

When an object appears red, which wavelengths of light are being **absorbed?** ______________________

______________________

When an object appears red, which wavelengths of light are being **reflected?** ______________________

______________________

## Experiment 1

The following experiment is designed to answer the following two questions: 1. How many pigments are in spinach juice? 2. Do all these pigments absorb the same part of the UV spectrum?

# Question 1—Paper Chromatography of Chloroplast Pigments

**Pigments** are any substances that absorb light. **Chlorophylls, xanthophylls and carotenes** are the major pigments found in the leaves of green plants. In this activity, you will use **paper chromatography** to separate the pigments present in an extract of spinach leaves. Paper chromatography is a relatively quick analytical chemistry technique used for separating and identifying mixtures of pigments.

Each table should obtain a piece of chromatography paper, taking care to touch only the edges of the paper (oils from hands will interfere with the separation). Place the chromatography paper on a clean sheet of paper. Use a lead pencil to draw a line across the paper 15 mm from the bottom, as illustrated.

15 mm

With a paint brush, apply a line of pigment extract on top of the pencil line that you drew. Let the pigment extract dry (drying keeps the pigment line from spreading too much). Repeat this process two more times until you have a total of 3 lines of pigment extract painted on top of the pencil line. When this is dry, write your initials on the top right corner of the chromatography paper and bring it to your instructor at the fume hood.

### *Important Safety Tip*

We are working in the fume hood because the solvent contains acetone and ether, both of which are highly flammable. Also we don't want to breathe ether—it was once used as a general anesthetic!

In a large beaker, your instructor will pour **solvent consisting of 10% acetone in petroleum ether.** With the green line on the outside, the chromatography paper will be placed into the beaker so that the paper touches the solvent but the green line does **NOT** touch the solvent (touching the green line with the solvent will prevent the separation of pigments into individual components because the pigments will travel rapidly together up the strip).

Watch as the solvent moves up the paper. Describe what you observed.

When the solvent has reached within a centimeter of the top of the paper, remove the paper from the beaker and let it dry in the fume hood. Return to your table to record your results.

** **While you wait for the solvent to finish moving up the paper and then to dry, go on to *Activity 2*** on page 144 of **this lab.**

Using colored pencils, sketch your results in the box provided, showing the relative position of the pigments and labeling the 4 pigments.

## What Colors Are the Pigments

To help you identify the pigments: chlorophyll *a* is blue-green, chlorophyll *b* is yellow-green, carotene is deep yellow, xanthophyll is pale yellow.

**Why did this happen?** Paper chromatography separates pigments on paper based on their solubility in the solvent. Compounds which are very soluble in the solvent move along with the advancing solvent front while less soluble pigments travel more slowly and remain near the bottom of the paper.

**Which pigment was most soluble?**

**Which pigment was least soluble?**

Once you have recorded your data from your chromatography strips, your instructor may collect and use these strips for the extraction of pigments, as follows. (OR, students may perform this portion of the experiment, if assigned by your instructor.)

## Question 2—Absorption of Light by Various Pigments

The term **absorption** refers to the physical process of absorbing light. We will collect data for ***absorbance***, which refers to the mathematical quantity of light that is absorbed.

Our first step will be the **extraction** of pigments. By this process, we will remove the pigments from the chromatography paper with an organic solvent, **acetone**, in which the four pigments are soluble.

a. Label **four** small beakers as: **chlorophyll *a*, chlorophyll *b*, carotene, xanthophyll.**

b. Using scissors, cut across the chromatography strip parallel to the bands of color, being careful to separate each pigment. Remove any excess white paper.

c. Place the strips for each pigment from all of the student groups into the appropriate beakers. You may want to cut these long strips into shorter segments to help them fit.

d. Pour 5 mL of **acetone** into each beaker. Swirl the contents of each beaker to wet the paper, then cover with a small piece of plastic wrap and wait for 5 minutes until all of the pigment has been removed from the paper.

e. Label four cuvettes (these are small test tubes used in the spectrophotometer) as follows: A (for chlorophyll a), B (for chlorophyll b), C (for carotene) and X (for xanthophyll). Use a lead pencil to label the cuvettes. Carefully pour the extracted pigments into the appropriate cuvette. Cuvettes should contain similar volumes of pigment extract to within 1 cm of the bottom of the white line.

f. Label a fifth cuvette as BLK for blank. The blank will "zero out" or calibrate the machine so that only the absorbance of each pigment is displayed (light will still be absorbed by the solvent; however, it will be subtracted from each reading).

What substance will be used for the blank? ______________________________

Our second step will be to collect data for **absorbance** of different wavelengths of light for each pigment. Using a spectrophotometer (or "spec" for short), we will shine different wavelengths of light through each of the four pigments to measure the absorbance of that particular wavelength of light. Let's quickly review how to use the spec and describe how we will collect data for this lab:

Your instructor will direct you as to which pigments your table will collect data for – generally each table will collect data for one pigment. Each team will record their data in a table for all students in the class to use in the next activity. Copy your peers' data into the table on the next page of the lab manual.

1. **Turn on the machine,** if this has not already been done and allow it to warm up for 10 minutes.
2. Following the instructions on the digital display, **press any key.** The machine will "buzz and whir" as it performs internal diagnostics and prepares to collect data.
3. With the lid to the sample chamber closed, **the digital display should read a wavelength in nm (nanometers) and give a number for A ( = absorbance).** If the data are not provided as A, toggle through the A/T/C key until they are.
4. Use the ▲ nm or ▼ nm to **set the wavelength to 400 nm** = the first wavelength for data collection.
5. Wipe the BLANK with a Kimwipe (to remove any dirt or oils from fingerprints), then **insert the BLANK into the sample chamber,** being sure to align the white mark with the front of the chamber. Close the lid to the sample chamber and **press the 0 ABS/100% T key.**
6. When the digital display reads 0 ( = zero) for absorbance, the spec is calibrated for this wavelength.
7. **Insert the cuvette containing the first pigment** into the sample chamber. Without touching any keys on the spec, read and record the value for A in your data table.
8. Use the ▲ nm or ▼ nm to **set the wavelength to 420 nm** = the next wavelength for data collection. Repeat Steps 5–7 for each subsequent wavelength.

**While waiting to use the spec, go on to *Activity* 2 of this lab.**

## Absorbance of Different Wavelengths of Light by Chloroplast Pigments

**Record your data for absorbance to the nearest tenth—absorbance is a unitless number since it is a ratio of light absorbed compared with the light transmitted through a sample.**

| Color (ROY G BIV) | Wavelength (nm) | Chlorophyll *a* | Chlorophyll *b* | Carotene | Xanthophyll |
|---|---|---|---|---|---|
| | 400 | | | | |
| | 420 | | | | |
| | 440 | | | | |
| | 460 | | | | |
| | 480 | | | | |
| | 500 | | | | |
| | 520 | | | | |
| | 540 | | | | |
| | 560 | | | | |
| | 580 | | | | |
| | 600 | | | | |
| | 620 | | | | |
| | 640 | | | | |
| | 660 | | | | |
| | 680 | | | | |
| | 700 | | | | |

Our third and final step will be to prepare a graph demonstrating **the relationship between absorbance and wavelength for chloroplast pigments.**

When preparing a graph, the **independent variable** is graphed on the x-axis. It is the variable that you have control over—you modify it to examine its effect on the dependent variable.

**Which variable is the INDEPENDENT variable in this experiment?** ______________________________

The **dependent variable** is graphed on the y-axis. It is what you measure in the experiment (= the data that you collect)—it is what is affected during your experiment.

**Which variable is the DEPENDENT variable in this experiment?** ______________________________

**Your graph will need a title and both axes should be labeled with units** (where appropriate). You will **use a different symbol or color for each pigment** so your graph will have 4 lines when you are done. Use a lead pencil to draw your graph and then add color only after you have checked it carefully for graphing errors.

**Below the x-axis, use colored pencils to add the colors for the wavelengths of visible light.**

**Title:** ______________________________

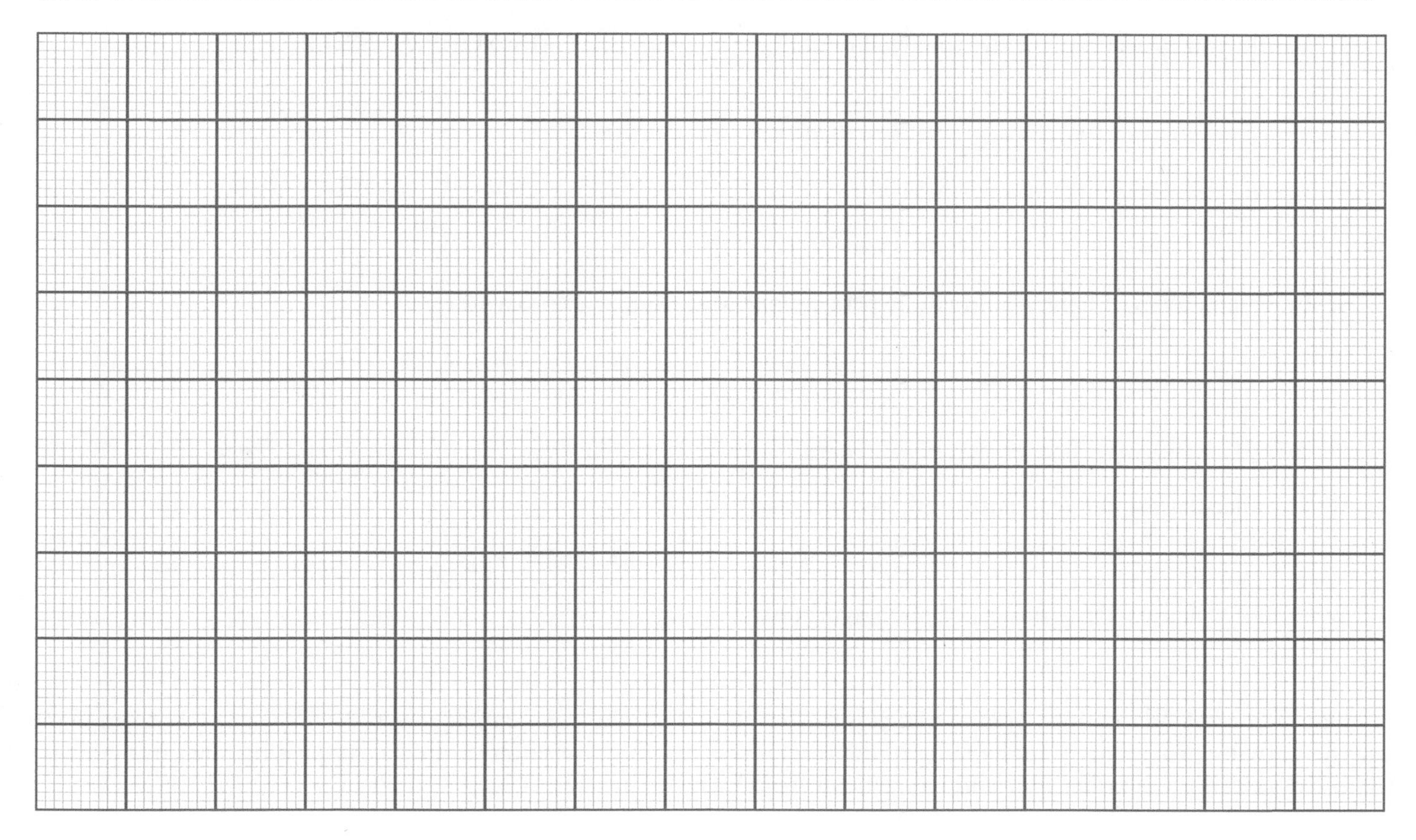

**Based on your graph, answer the following questions.**

1. At what wavelengths and colors did **chlorophyll *a*** absorb energy (*i.e.*, when was absorbance high?)

______________________________

______________________________

2. At what wavelengths and colors did **chlorophyll b** absorb energy?

______________________________

3. At what wavelengths and colors did **carotene** absorb energy?

______________________________

4. At what wavelengths and colors did **xanthophyll** absorb energy?

______________________________

5. Why do green plants appear green?

______________________________

6. **Only chlorophyll provides energy for the light reactions in photosynthesis.** So what is the purpose of the other pigments that were masked by chlorophyll in the spinach leaves? Compare your answers to Questions 1 & 2 with Questions 3 & 4 to help you answer this!!

7. Would you expect a plant to grow well in only green light? EXPLAIN YOUR ANSWER!!

8. In which colors of light would you expect a plant to maximize photosynthesis?

## Activity 2—The Reactants and Products of Photosynthesis

What molecules are the **reactants** of photosynthesis?

What molecules are the **products** of photosynthesis?

One of the key adaptations of land plants is the development of a waxy **cuticle** that prevents water loss from plants (see the drawing provided). Actually none of the reactants or products of photosynthesis can pass through this cuticle. **So how does carbon dioxide enter and oxygen exit from leaves?**

In addition to the products in the summary reaction of photosynthesis, plants lose water from their leaves in a process known as **transpiration.** Transpiration is an important process for plants as it pulls water up from the roots to the leaves through vessels known as **xylem. So how does water vapor exit from leaves?**

What are the openings on the lower side of this leaf called?

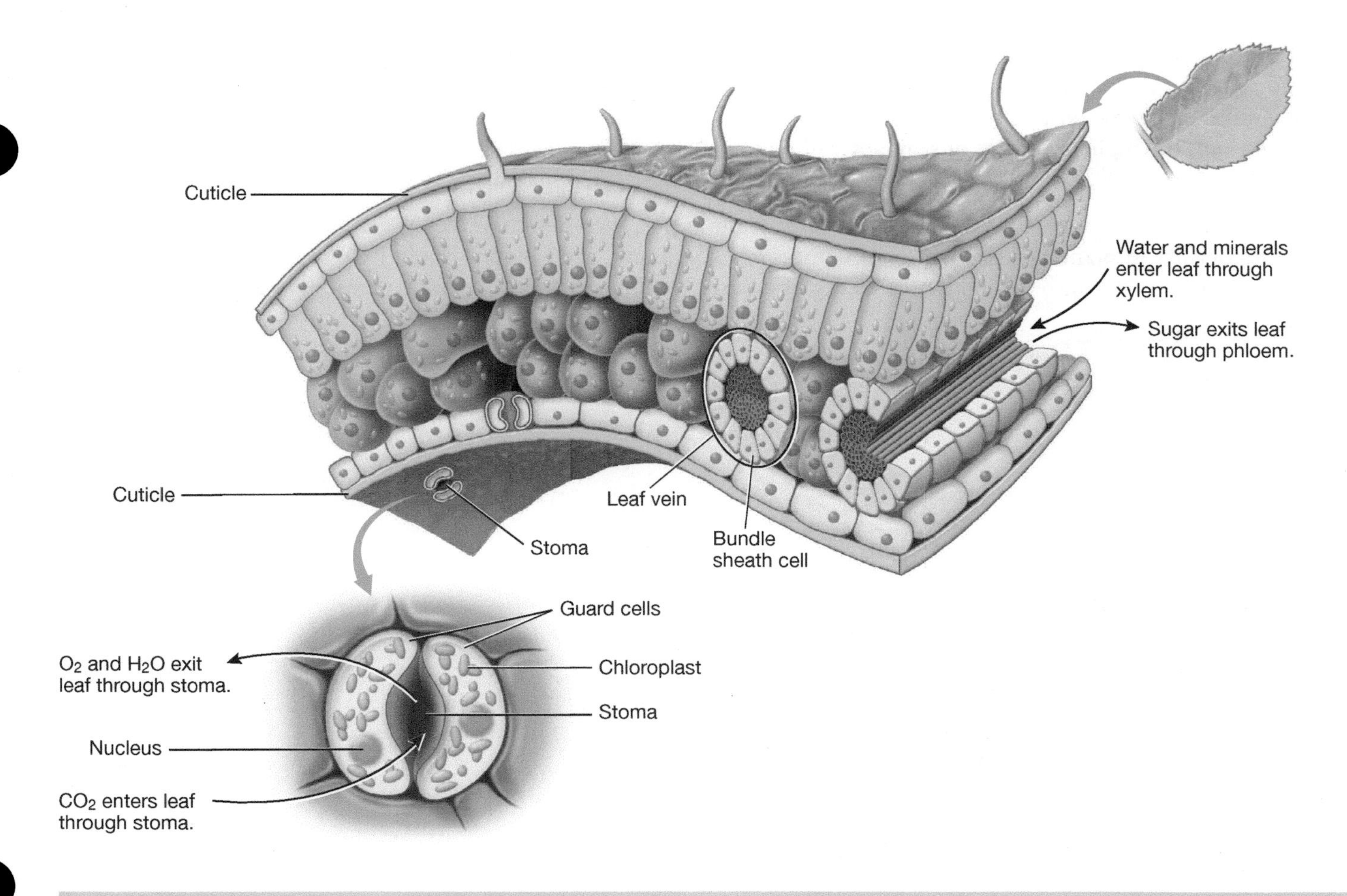


## Activity 3—Studying Stomata

**Stomata (stoma = singular;** Greek for *opening*) are the holes on plant leaves that allow carbon dioxide to enter and oxygen and water vapor to exit from leaves. Guard cells that flank the stoma open and close to regulate the passage of materials. When these cells fill up with water, they bulge and result in the opening of the stoma.

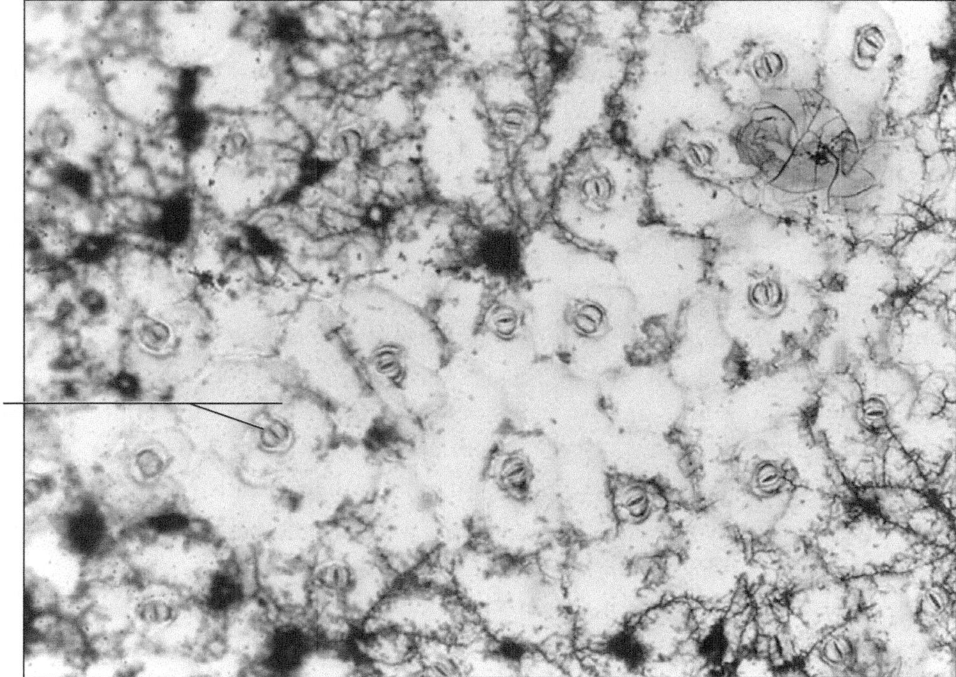

In this part of today's lab, you will prepare **stomata impressions** and examine how the density of stomata varies between the upper side and the underside of leaves. Note that the more stomata per unit area (stomata density), the more $CO_2$ can be taken up and the more water can be released. Plant leaves may have as many as 100,000 stomata per square cm. **Where do you think the highest density of stomata will be found?**

We will use the scientific method to answer this question!!

Working as a group at your table, observe stomata in plant leaves, then design and conduct an experiment that studies stomata density in plants' leaves. Before designing an experiment, each group should first have a question to answer and then develop a testable hypothesis. Next, design an experiment to answer your question.

Title for your experiment: ______________________

Question ______________________

Hypothesis ______________________

Independent variable(s) ______________________

Dependent variable(s) ______________________

Control variables ______________________

Explain how you will conduct the experiment ______________________

______________________

______________________

______________________

______________________

## Conduct the experiment

Method for Obtaining Stomata Impressions [Modified from 2004–Bruce W. Grant, Itzick Vatnick, and the Ecological Society of America. TIEE–Volume 1 ©2004–Ecological Society of America. Teaching Issues and Experiments in Ecology (TIEE) is a project of the Education and Human Resources Committee of the Ecological Society of America.]

1. Obtain the leaf upon which you wish to count stomata. Place the leaf on a piece of white paper.
2. On the side you wish to census stomata, paint a **thick swath** of clear nail polish.
3. After the nail polish has dried (let stand 5 minutes in the fume hood), obtain a square of VERY CLEAR tape (such as package sealing tape, but do NOT use scotch tape). Stick your tape piece to the area that contains the dried nail polish swath.
4. GENTLY, peel your nail polish swath from the leaf completely. You will see a cloudy impression of the leaf surface now attached to your tape piece (hereafter referred to as your "leaf impression").
5. Tape your leaf impression to a VERY CLEAN slide and use scissors to cut off the excess tape.
6. Use a china marker to identify the slide for the top and the slide for the bottom of the leaf.
7. Focus your leaf impression under 100× total magnification and observe the stomata.
8. Search around on your impression to find an area that subjectively appears to have a high density of stomata. That is, move the slide around until the field of view is away from the edge of the impression and so that there are no dirt blobs, no thumbprints, no damaged areas, and no big leaf vein impressions in view.
9. Count all stomata you see in this field of view and record data in the table provided.
10. Repeat the previous two steps three times, and calculate the average number (add the three counts together and divide by three) – this will be your one datum from this impression.
11. Repeat all steps above for leaf impressions in each treatment group.

Examine and draw your stomata impression at 100×.
Label stoma, guard cells.

Record the results in a table.

Number of stomata

| Top of leaf | Bottom of leaf |
|---|---|
| | |
| | |
| | |
| | |
| Mean = | Mean = |

Discuss your findings.

Did your data support your hypothesis? (Yes or No is not a complete answer to this question!)

## Thought questions:

1. Why might it be adaptive for stomata to occur mostly (if not entirely) on the undersides of leaves?

2. What plants would you expect to show the reverse pattern with stomata only on the upper leaf surface?

# Review Questions

**1** Write a balanced summary equation for photosynthesis.

**2** Complete the following sentences to trace the movement of compounds involved in photosynthesis. When choices of words are provided, circle the correct word. Fill in all blanks.

Carbon dioxide enters/exits the leaf by ______.

Oxygen enters/exits the leaf by ______.

Water enters the leaf by ______.

Plants lose water by the process known as ______ and this water vapor exits the leaf by ______.

The sugar produced by photosynthesis is ______ and it is stored as the polysaccharide ______ in leaves and other parts of plants.

**3** Key terms to define.

a. pigment ______.

b. chloroplast ______.

c. chlorophyll ______.

d. absorption ______.

e. absorbance ______.

f. cuticle ______.

g. stomata ______.

**4** Which wavelengths of light provide the energy for photosynthesis?

a. Those that are reflected.

b. Those that are absorbed.

**5** Which wavelengths of light produce the colors that we see?

a. Those that are reflected.

b. Those that are absorbed.

# The Green Machine I
## Understanding Basal Plants

# 11

*Every life form on the surface of this planet is here because a plant was able to gather sunlight and store it, and something else was able to eat that plant and take that sunlight energy in to power its body.* —**Thom Hartmann (1951–present)**

## OBJECTIVES

*At the completion of this chapter, the student will be able to:*

1. Describe basic plant tissues.
2. Discuss the importance of and origin of plants.
3. Compare and contrast nonvascular and vascular plants.
4. Describe the process of alternation of generations.
5. State the characteristics and provide examples of nonvascular plants.
6. Compare and contrast reproduction and other characteristics of seedless vascular plants and seed plants.
7. Discuss the functions of xylem, phloem, roots, stems, and leaves.
8. Describe the basic biology and life cycle of ferns.

## Just wondering . . .

*Answer the following questions prior to coming to lab.*

1. Are there any plants that have a parasitic lifestyle?

2. Are ferns commercially important?

3. What is the relationship between a seedless vascular plant known as *Lycopodium* and the history of photography?

4. What was the world of ancient plants like during the Carboniferous period?

5. What evolutionary adaptations helped ancient plants colonize the land?

A wise man once declared, "We need the plants much more than they need us." Have you ever stopped to thank a plant? Think about it! Plants provide oxygen, food, shelter, shade, erosion control, and commercial products for human uses, such as timber, medicine, and even the paper you are looking at right now. In addition, many species of plants are aesthetically pleasing such as roses. Remember—the next time you are sitting under a majestic tree—say thank you!

Plants are a diverse group of eukaryotic, multicellular, photosynthetic autotrophs that inhabit myriad environments from lush tropical rainforests to scorching deserts. Figure 11.1 is a cladogram of the diverse world of plants. Biologists have identified over 300,000 species of plants and estimate 400,000 species may exist. Plants vary in size from the smallest flowering plant, *Wolffia angusta* (a duckweed), measuring less than 1 mm in diameter, to the giant *Sequoia sempervirens*, which measures nearly 120 m tall. The oldest plant in the world is thought to be the King Holly (*Lomatia tasmanica*), a shrub that lives in Tasmania. This remarkable plant is estimated to be more than 43,000 years old!

Paleobotanists believe plants evolved during the Paleozoic era from freshwater green algae known as charophytes, approximately 450 million years ago. To make the transition from the aquatic environment, ancestral plants had to evolve mechanisms that prevent desiccation, anchor the plant body, transport water and nutrients, and ensure propagation of the species.

In this chapter, we will look at plant histology and then dive into the nonvascular and seedless vascular plants.

## Plant Histology

Plants being multicellular organisms are comprised of a number of specialized tissues. The study of tissues is termed *histology*. Histology complements the study of gross anatomy and provides the structural basis for studying organs and organ systems.

### Meristematic Tissue

Plants have permanent regions of growth composed of **meristematic tissues.** In these tissues, cells are actively undergoing cell division. The new cells resulting from cell division usually are small and six-sided with a prominent nucleus. As the newly formed cells mature, they begin to take on their characteristic size, shape, and function.

Meristematic tissues found at or near the tips of roots and stems make up the **apical meristem.** Growth of the apical meristem, known as primary growth, involves increasing the length of the root or stem. The apical meristem gives rise to three distinct regions:

1. the **protoderm,** which gives rise to the epidermis
2. the **ground meristem,** which gives rise to building block tissue called parenchyma that usually exists between the epidermis and the vascular tissue
3. the **procambium,** which gives rise to vascular tissue such as xylem and phloem

The lateral meristem provides the plant growth in girth, or secondary growth. Two derivatives of the lateral meristem are:

1. the **vascular cambium,** or simply cambium, which gives rise to tissues important in support and protection
2. the **cork cambium** in woody plants, which gives rise to cork tissue that makes up the protective bark; the cork is impregnated with the waxy substance suberin, which makes the cells impenetrable to water

Grasses do not possess a vascular cambium or cork cambium, but they do have apical meristematic tissue called intercalary meristems near nodes (regions of leaf attachment) at intervals throughout the plant. Intercalary meristems allow grass to grow back quickly after being grazed by a cow or cut by a lawn mower.

### Parenchyma Tissue

**Parenchyma tissue** is composed primarily of parenchyma cells, the most abundant and diverse type of cell in plants. Parenchyma cells vary in size and shape and tend to have large vacuoles. Parenchyma cells are involved in storage, photosynthesis, support, secretion, repair, and the movement of water and food in plants.

The soft, edible parts of apples and other fruits consist mostly of parenchyma cells. In potatoes, parenchyma cells store starch. Parenchyma cells with numerous chloroplasts are sites of photosynthesis. These cells, chlorenchyma, are abundant in leaves and stems of herbaceous plants. Parenchyma cells with extensive air spaces found in water plants are known as aerenchyma. This tissue helps to support the plant and, when squeezed, is crunchy.

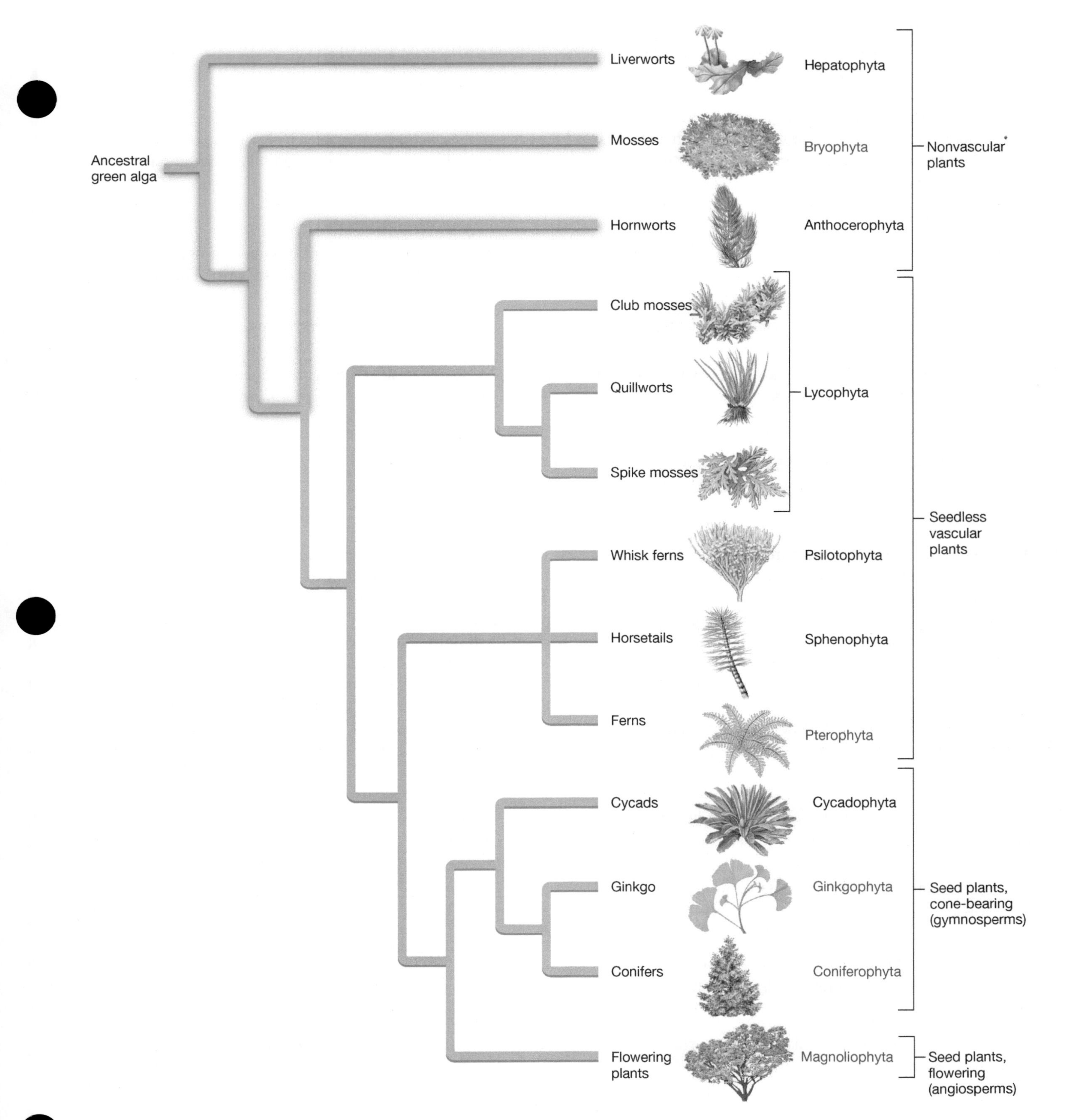


FIGURE **11.1** Phylogenetic relationships and classification of Plantae. The phyla shown in red are discussed in this and the following chapter.

Some mature parenchyma cells can divide when stimulated. When a plant is damaged, parenchyma cells are important in repair. When gardeners make cuttings, they take advantage of the growth of parenchyma cells.

## Collenchyma Tissue

**Collenchyma tissue** is composed of elongated collenchyma cells. These cells develop thick, flexible walls that support young plants and specific plant structures such as leaves and flower parts. Collenchyma resides beneath the epidermis in stems. In a fresh specimen, collenchyma tissue glistens. This tissue borders the veins of leaves and makes up the "strings" of a stalk of celery.

## Sclerenchyma Tissue

**Sclerenchyma tissue** is composed of thick sclerenchyma cells, often impregnated with the plant polymer **lignin.** Unlike parenchyma and collenchyma cells, sclerenchyma cells are dead at maturity and primarily provide support. The two types of sclerenchyma are:

1. **fibers,** the long, slender cells that occur in strands. They are commonly found in roots, stems, leaves, and fruits. Fibers are used in the manufacture of ropes, string, and canvas.
2. **sclereids,** or stone cells, are responsible for the gritty texture of plants, in which they may occur singly or in groups throughout; also, a major component of the shell of various nuts and the pit of a peach.

Complex tissues consist of two or more types of cell. Complex tissues can be divided into:

1. **dermal tissue,** consisting of the epidermis and the periderm
2. **vascular tissues,** consisting of xylem and phloem

## Epidermal Tissue

The epidermis constitutes the outermost layer of cells in plant structures, such as roots, stems, leaves, floral parts, fruits, and seeds. The epidermis generally is one cell layer thick and does not undergo photosynthesis. Because epidermal cells are in direct contact with the environment, they vary in form and function. The walls of many epidermal cells are covered with a waxy cuticle, minimizing water loss and protecting the plant against pathogens. The waxy cuticle can be easily observed on magnolia leaves. Epidermal cells also can form **root** and **leaf hairs** that increase the surface area. Numerous small, pore-like structures are found primarily on the underside of leaves. The structures are called **stomata** (singular, stoma), and they are bordered by a pair of **guard cells.** Stomata allow for gas exchange in plants.

In the roots and stems of woody plants, the epidermis is sloughed off and replaced by the periderm, which makes up the outer bark composed of box-shaped cork cells. Mature cork cells are dead. The fatty substance suberin is found in the walls of cork cells, providing protection from mechanical injury, desiccation, and extreme temperatures.

## Vascular Tissue

Vascular tissue makes up the **vascular bundle.** In plants, xylem tissue is the primary water-conducting tissue and also serves in support, food storage, and the conduction of minerals. Xylem is composed of these basic types of cells:

- parenchyma cells serve in storage
- tracheids and vessel elements are the major conducting cells of the xylem
- ray cells serve in lateral conduction and storage
- fibers also can occur in xylem, adding support and storage

Phloem tissue conducts dissolved food materials throughout the plant body. The food is composed primarily of sugars produced through photosynthesis. Phloem is composed of:

- parenchyma cells, which provide storage
- sclerenchyma, which provide support
- sieve tube members, which provide conduction
- companion cells, which also provide conduction

### Hints & Tips

1. Read the description of the tissue, and study the pictures thoroughly.
2. Do not become dependent on the color of the tissue.
3. View the tissue using the microscope powers suggested by your instructor.
4. In viewing, scan different parts of the slide and use different depths of field.
5. Accurately draw and color what you view.
6. Spend quality time viewing the specimens. *Do not rush!*

# Plant Classification

The absence or presence of specialized conducting tissue known as **vascular tissue** is a common way to distinguish plants. **Nonvascular plants** lack specialized conducting tissues to transport water and nutrients throughout the plant's body. In addition, these plants lack true roots, stems, and leaves. Examples of nonvascular plants are liverworts, hornworts, and true mosses. Presently, about 25,000 species of nonvascular plants have been identified. Most living plants are considered vascular plants because they possess an extensive conducting system composed of specialized tissues.

Vascular plants can be divided into the seedless vascular plants and the seed plants. The **seedless vascular plants** include the club mosses and the ferns. The **seed plants** are the largest group of vascular plants and include gymnosperms (such as ginkgos, cycads, and conifers) and angiosperms (flowering plants, such as zinnias and elms).

The life cycle of plants is characterized by an **alternation of generations** (Fig. 11.2). In this process, two distinct generations give rise to each other. The haploid (*n*) **gametophyte** generation is characterized by the production of male and female gametes through mitosis. The male and female gametes fuse during fertilization, forming a diploid **sporophyte.** The sporophyte generation is diploid (2*n*). It produces haploid spores that undergo mitosis to form a gametophyte. In nonvascular plants the gametophyte generation is dominant, but in seedless vascular plants and seed plants, the sporophyte generation dominates. A fern, a pine tree, and a tulip are all examples of sporophytes.

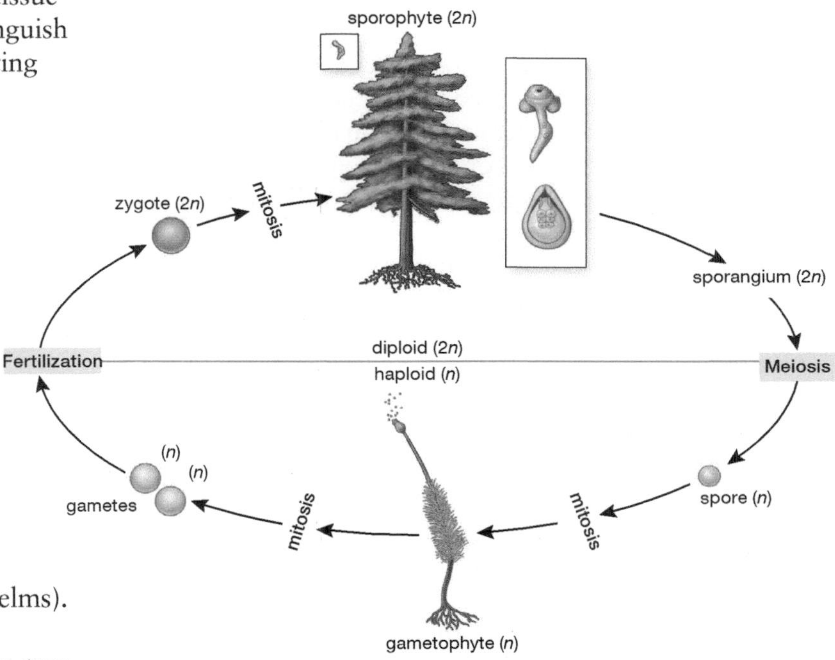


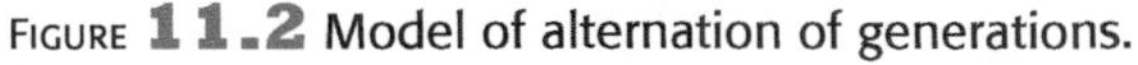
FIGURE **11.2** Model of alternation of generations.

# Nonvascular Plants

In the history of plants on Earth, the nonvascular plants played an important role in establishing the transition from water to land and dominance of the gametophyte generation. The nonvascular plants are generally small and herbaceous (nonwoody). Although most of these plants are found in moist environments, some species can survive in arid environments. Three distinct phyla of nonvascular plants (Fig. 11.3) have been established:

1. Phylum Hepatophyta, the liverworts
2. Phylum Anthocerophyta, the hornworts
3. Phylum Bryophyta, the true mosses*

* The term *bryophyte* is often used to describe all of the nonvascular plants, even though it is the proper taxonomic name for the phylum in which the true mosses are classified. Only phylum Bryophyta will be represented in this chapter.

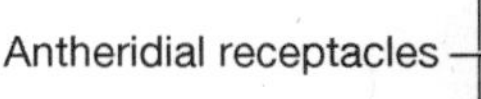


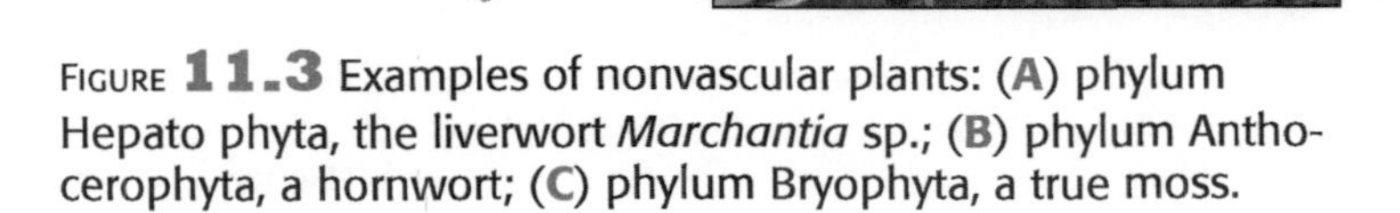
FIGURE **11.3** Examples of nonvascular plants: (**A**) phylum Hepato phyta, the liverwort *Marchantia* sp.; (**B**) phylum Anthocerophyta, a hornwort; (**C**) phylum Bryophyta, a true moss.

# EXERCISE 11.1 Phylum Bryophyta

True mosses have been placed in phylum Bryophyta. Presently, nearly 15,000 species of mosses have been identified, but many small plants called mosses do not fall into this group. For instance, Spanish moss is actually a member of the pineapple family, reindeer moss is actually a lichen, Irish moss is a red algae, and club mosses are lower vascular plants.

The gametophyte stage of mosses consists of small, spirally arranged, leaflike structures surrounding a central axis. The blades of the leaflike structure are one cell layer thick, lack vascular tissue and stomata, and surround a thickened midrib. Rhizoids anchor mosses to their substrate.

Mosses are capable of asexual reproduction through fragmentation, but they undergo an alternation of generation with gametophyte and sporophyte stages. The "leafy" gametophytes are male, bearing antheridia, or female, bearing archegonia. Flagellated sperm cells exit the antheridia and travel with the aid of water to the archegonia, where a single egg is fertilized. The zygote undergoes meiosis, forming spores housed in the sporophyte. The sporophyte appears as a tall stalk topped by a **sporangium**, or capsule. The calyptra protects the capsule. A foot connects the seta, or stalk, of the sporophyte to the leafy gametophyte. In some mosses, a single capsule may contain 50 million spores.

The tip of the capsule consists of a lid-like structure called the **operculum**. Teeth-like structures form the **peristome** (which means around the mouth) and lock the operculum to the capsule. During dry conditions, they unlock and allow spores to be carried by the wind. Through a hand lens, the peristome appears ornate and usually is orange or red. Immature spores land on a substrate and under suitable conditions develop into a filamentous protonema. The protonema eventually develops into the "leafy" gametophyte, and the cycle begins again (Fig. 11.4).

Most mosses live in moist environments of the temperate zone, although several species have been found living in the Arctic, Antarctic, and even deserts. Frequently, mosses can be seen growing on the trunks of trees and on the sides of buildings. After a fire or volcano, some mosses serve as pioneer species and help form soil.

Commercially, one of the most valuable species of moss is *Sphagnum*, or peat moss. In many regions, deposits of peat are mined for their use as packing material and fuel. Peat can absorb large amounts of water and is used in the gardening industry to enhance the water-holding capacity of soil and potted plants.

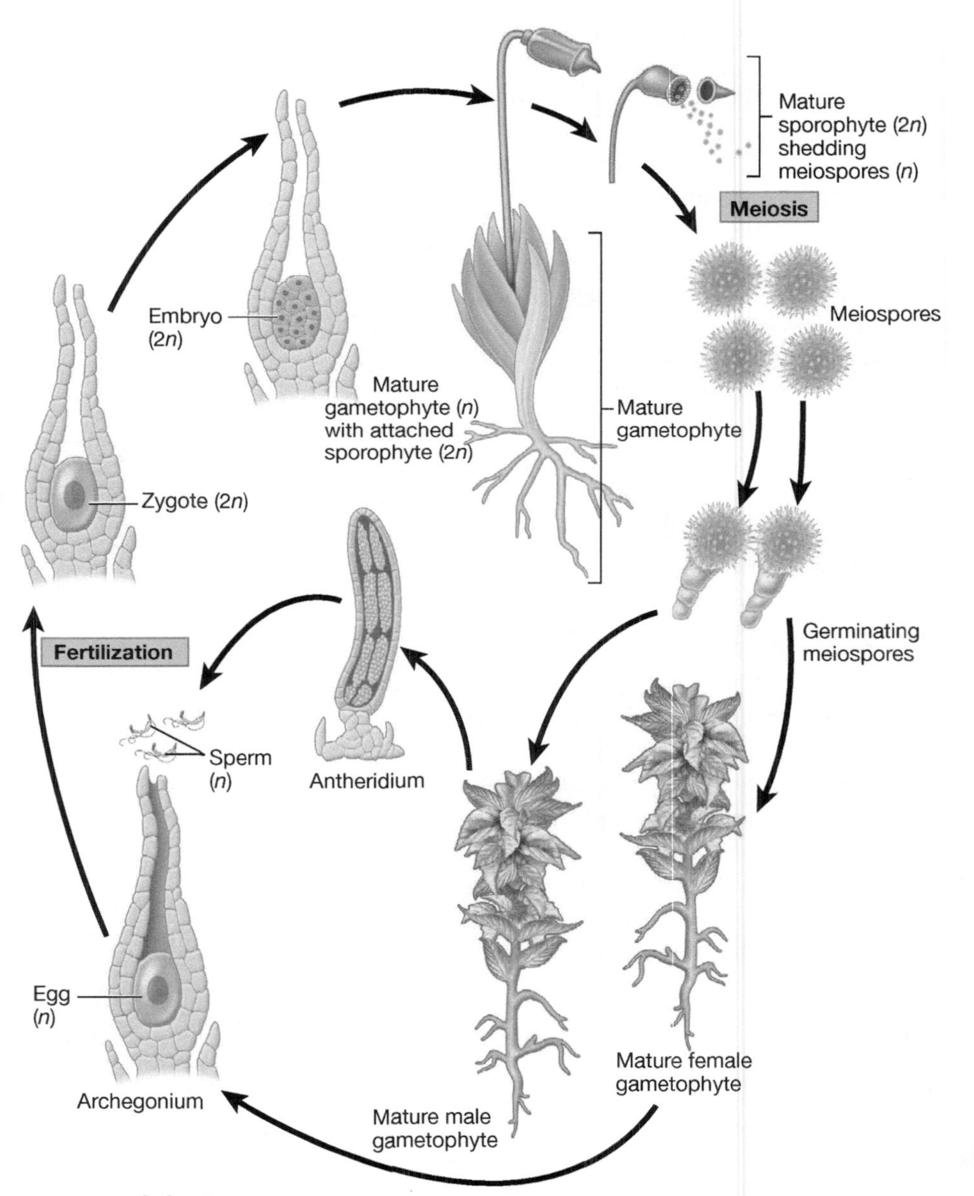


FIGURE 11.4 Life cycle of a moss (Bryophyta).

## Did you know . . .

### Peat Moss in History

The use of peat moss, or *Sphagnum*, has been mentioned in folklore as far back as the 11th century. Native Americans used peat moss in diapers, and some ancient societies used it in menstrual pads. Medical texts in the 1800s mentioned that *Sphagnum* could be used as bandages or to pack abscesses. During the Russo-Japanese War from 1904 to 1905, *Sphagnum* was commonly used as bandage material. During World War I, in a time of intense fighting in France, hospital staffs resorted to packing wounds and making bandages with *Sphagnum*. To their surprise, not only was the peat moss more absorbent than cotton but also the infection rate among the wounded was reduced significantly. The remarkable peat moss was used in World War II in a similar fashion. Peat moss has antibacterial properties as the result of the presence of phenol compounds, low pH, and the polysaccharide chitosan.

The moss *Sphagnum* was used as bandages during World War I.

## Procedure 1

### Macroanatomy of *Polytrichum*

**Materials**

- ❑ Dissecting microscope
- ❑ Living specimen of *Polytrichum*
- ❑ Colored pencils
- ❑ Camera

*Polytrichum* sp., haircap moss, is a common moss found living in bogs. It is a relatively large moss, perhaps reaching 10 cm in length in certain environments. It is distinguished by having a tall sporangium with a golden calyptra sitting atop the seta and protecting the capsule (Fig. 11.5).

1. Procure the necessary supplies and equipment.
2. Using a dissecting microscope, draw and label the specimen of *Polytrichum*. Pay particular attention to the surface, rhizoids, and reproductive structures. Place your labeled sketches and observations below.

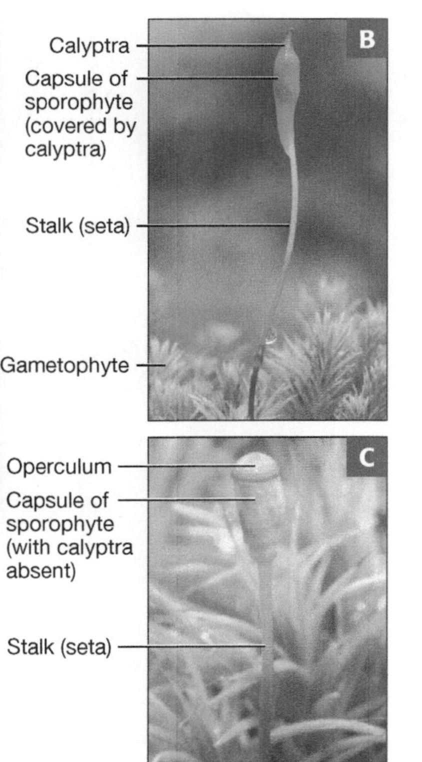



FIGURE **11.5** (**A**) *Polytrichum* sp., a common moss often used in coursework; (**B**) gametophyte plants with sporophyte plant attached; (**C**) sporophyte plant and capsule.

*Polytrichum*

## Mosses—Complete Drawing #2 in the Handout on Alternation of Generation in Plants after drawing these slides.

| | |
|---|---|
| Draw the **moss antheridium** from prepared slides of *Mnium* or *Polytrichium*. **What is produced in the antheridia**? | Draw the **moss archegonium** from prepared slides of *Mnium* or *Polytrichium*. **What is produced in the archegonium**? **How does the sperm reach the egg in mosses**? |
| Draw the **moss capsule** from prepared slides of *Mnium* or *Polytrichium*. **What is produced in the capsule**? | Draw the **moss protonema** from prepared slides of *Mnium* or *Polytrichium*. **What does the protonema develop from**? **What does the protonema develop into**? |

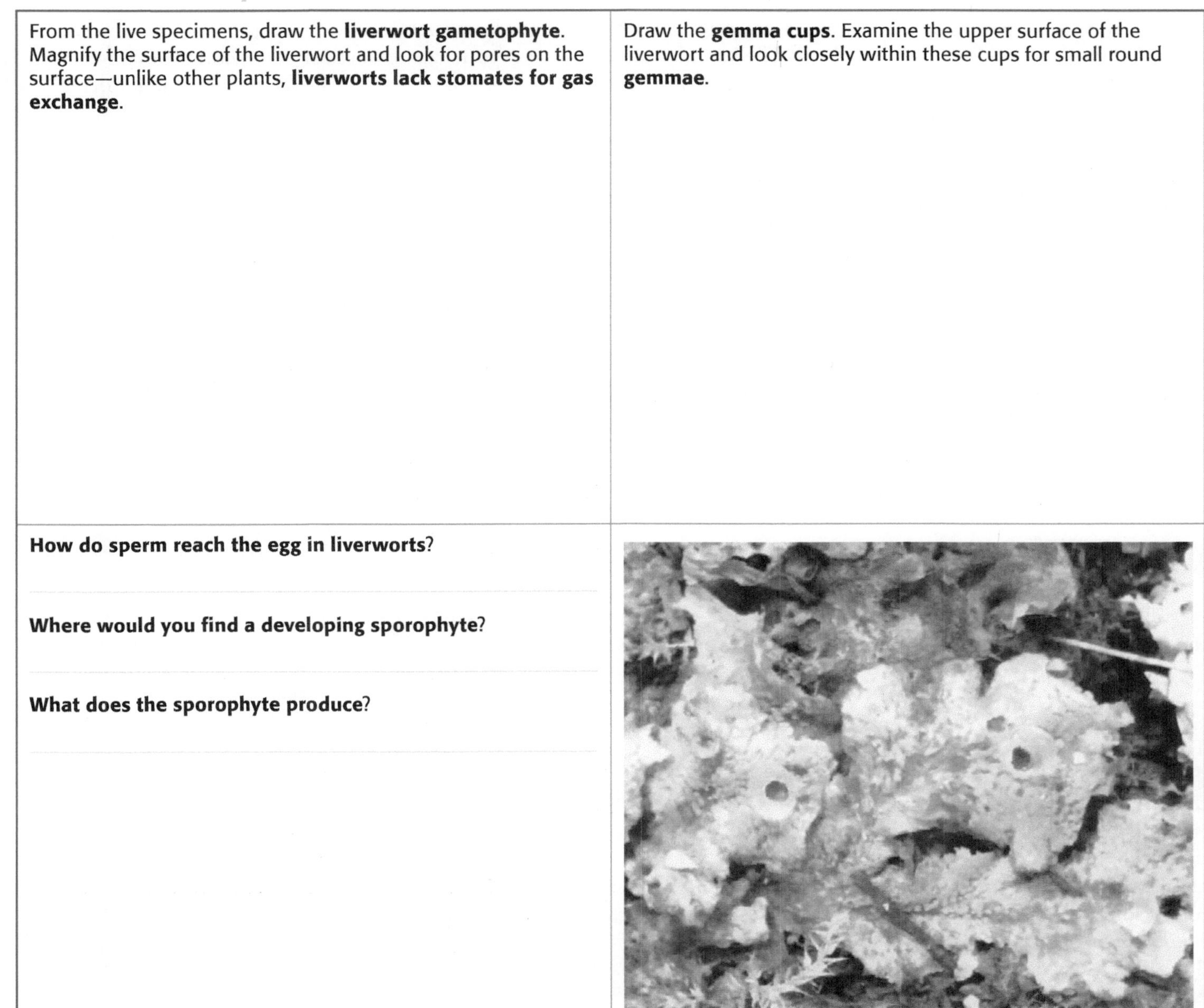

| From the live specimens, draw the **liverwort gametophyte**. Magnify the surface of the liverwort and look for pores on the surface—unlike other plants, **liverworts lack stomates for gas exchange**. | Draw the **gemma cups**. Examine the upper surface of the liverwort and look closely within these cups for small round **gemmae**. |
|---|---|
| **How do sperm reach the egg in liverworts**?<br><br>**Where would you find a developing sporophyte**?<br><br>**What does the sporophyte produce**? | |

Liverworts produce **gemmae** by mitosis as a means of asexual reproduction. Splashing raindrops or small animals help to disperse gemmae. When a gemma lands in a suitably moist environment, it germinates and grows into a new liverwort gametophyte.

# Vascular Plants

During the Silurian period 443–416 million years ago ancient vascular plants such as *Cooksonia* (Fig. 11.6) established the foundations of modern vascular plants. The vascular plants possess specialized tissues for conducting water and nutrients throughout the plant. In modern vascular plants, **xylem** conducts water and dissolved minerals, and **phloem** conducts nutrients, such as sucrose, hormones, and other molecules. In some plants the cells responsible for conducting water are often strengthened by the polymer lignin, which allows the plant to grow tall. The central portion of the roots and stems of vascular plants has a **stele** composed of xylem and phloem. Ancient vascular plants, such as lycophytes, possess a protostele in which a core of xylem is surrounded by phloem. More modern plants, such assunflowers, possess a siphonostele in which a central spongy pith is surrounded by a ring of xylem and phloem. Other plants, such as conifers and the majority offlowering plants, have eusteles with xylem and phloem arranged into vascular bundles. In vascular plants, the sporophyte generation is dominant and possesses the vascular tissues. Vascular plants have a waxy **cuticle** for protection againstdesiccation as well as small openings called **stomata** on photosynthetic structures to allow for gas exchange. Vascular plants also possess true roots, stems, and leaves:

1. Roots are plant organs that absorb water and nutrients from the soil and anchor a plant.
2. Stems are vascular plant organs that support leaves and reproductive structures.
3. Leaves are the primary photosynthetic organs of plants.

The two major types of vascular plants are the seedless vascular plants and the seed plants. The seed plants are covered in the next chapter.

FIGURE **11.6** *Cooksonia* sp. is one of the oldest vascular land plants.

## Seedless Vascular Plants

The seedless vascular plants, as the name indicates, do not produce seeds but instead reproduce through the production of spores like their ancestors. The seedless vascular plants dominated the landscape during the Devonian and Carboniferous periods. Unlike the relatively small seedless vascular plants of today, some species of seedless vascular plants, such as *Lepidodendron* (scale tree), reached heights exceeding 30 m. The vast coal deposits of the Carboniferous period are the result of carbonization of seedless vascular plants that resided in giant swamp forests. Today, the seedless vascular plants are represented by the following four phyla:

1. Lycophyta (club mosses)
2. Sphenophyta (horsetails)
3. Psilotophyta (whisk ferns)
4. Pterophyta (ferns).

Examples of each are shown in Figure 11.7.

FIGURE **11.7** Examples of seedless vascular plants: (A) phylum Lycophyta (club mosses); (B) phylum Sphenophyta (horsetails); (C) phylum Psilotophyta (whisk ferns); (D) phylum Pterophyta (ferns).

# EXERCISE 11.2 Phylum Pterophyta

Ferns placed in phylum Pterophyta (Polypodiophyta) are the most abundant seedless vascular plants. Most fern species live in moist, tropical regions of Earth, although some species reside in temperate regions as well as the Arctic Circle. Some ferns can even live in aquatic environments and dry areas. Ferns range in size from giant tropical tree ferns that can exceed 28 meters in height to *Azolla*, a diminutive aquatic fern that measures less than a centimeter in diameter (Fig. 11.8). Scientists believe ancestral ferns first appeared in the Devonian period approximately 375 million years ago. Today, plant taxonomists have identified approximately 11,000 species of ferns.

Ferns are cherished for their ornamental value. They are used for outdoor decorations and by florists to construct bouquets. The Environmental Protection Agency (EPA) suggests ferns are valuable in filtering formaldehyde and other toxins from the air. Fern rhizoids and fronds are foods in many cultures. Bracken fern fronds were used in the past to thatch roofs. Medicinally, ferns and their products have been used in the treatment of leprosy, parasitic worms, labor pains, sore throat, diabetes, dandruff, and many other maladies.

FIGURE **11.8** (A) Tree fern, *Cyathea* sp.; (B) aquatic fern, *Azolla* sp.

The leaves of ferns, known as **fronds**, arise from rhizomes. Immature fronds develop from the tip of a rhizome and appear as a tightly coiled and rolled-up structure called a **fiddlehead** (Fig. 11.9). Compound frond ferns possess ornate leaflets, or **pinnae** (Fig. 11.10). Simple frond ferns have leathery, broad, unbranched, strap-like fronds. The pinnae are attached to a midrib, sometimes called a rachis. A **petiole**, or stalk, attaches the pinnae to the rhizome. Branching roots also arise from the rhizomes.

The sporophyte stage is dominant in ferns. Most fern species are homosporous, producing one kind of spore. The spores are produced in sporangia, appearing as distinct brown spots on the underside of the frond, called **sori** (Fig. 11.11). To an untrained eye, the sori may appear like a fungus or insect eggs. In some species, the sori are protected by a colorless flap called an **indusium**. The **annulus**, a fuzzy region of the sorus, catapults the mature spores. Adder's tongue fern may produce more than 15,000 spores per sorus. The collective production of spores in some ferns exceeds 50 million. The water fern *Marsilea* and several other species are heterosporous, producing spores in a **sporocarp**. Spores that land in a favorable environment germinate, producing heart-shaped gametophytes, or **prothalli**. The gametophytes possess rhizoids that anchor them to their substrate. Flagellated sperm cells are produced in the antheridia of the gametophyte. The sperm are released and swim to the archegonium, where they fertilize the awaiting egg, forming a zygote. The zygote develops into the sporophyte generation, completing the life cycle of a fern (Fig. 11.12).

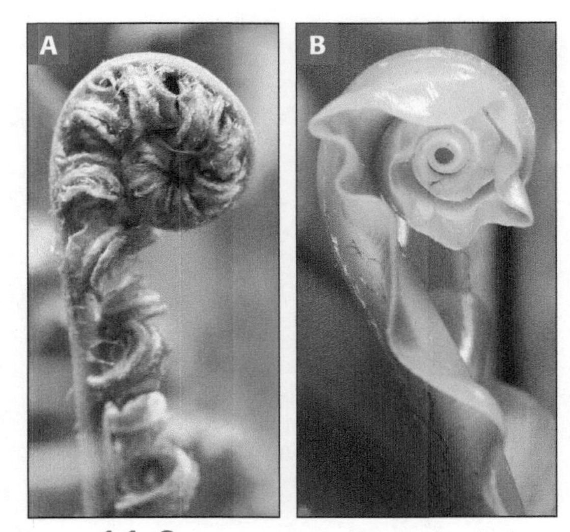

FIGURE **11.9** (A) Compound; (B) simple fern leaf forming a fiddlehead.

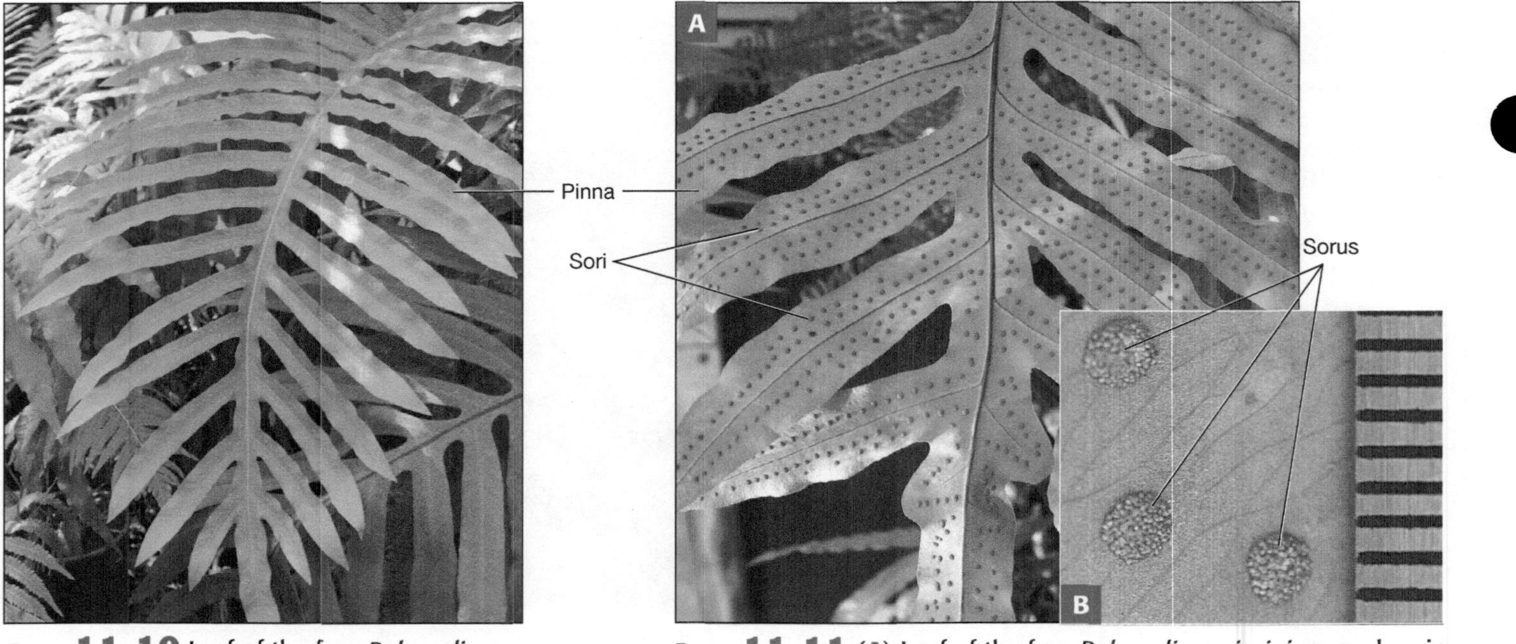


FIGURE **11.10** Leaf of the fern *Polypodium virginianum*.

FIGURE **11.11** (A) Leaf of the fern *Polypodium virginianum* showing sori (groups of sporangia); (B) closeup of the sori on the fern pinna of *Polypodium virginianum* (scale in mm).

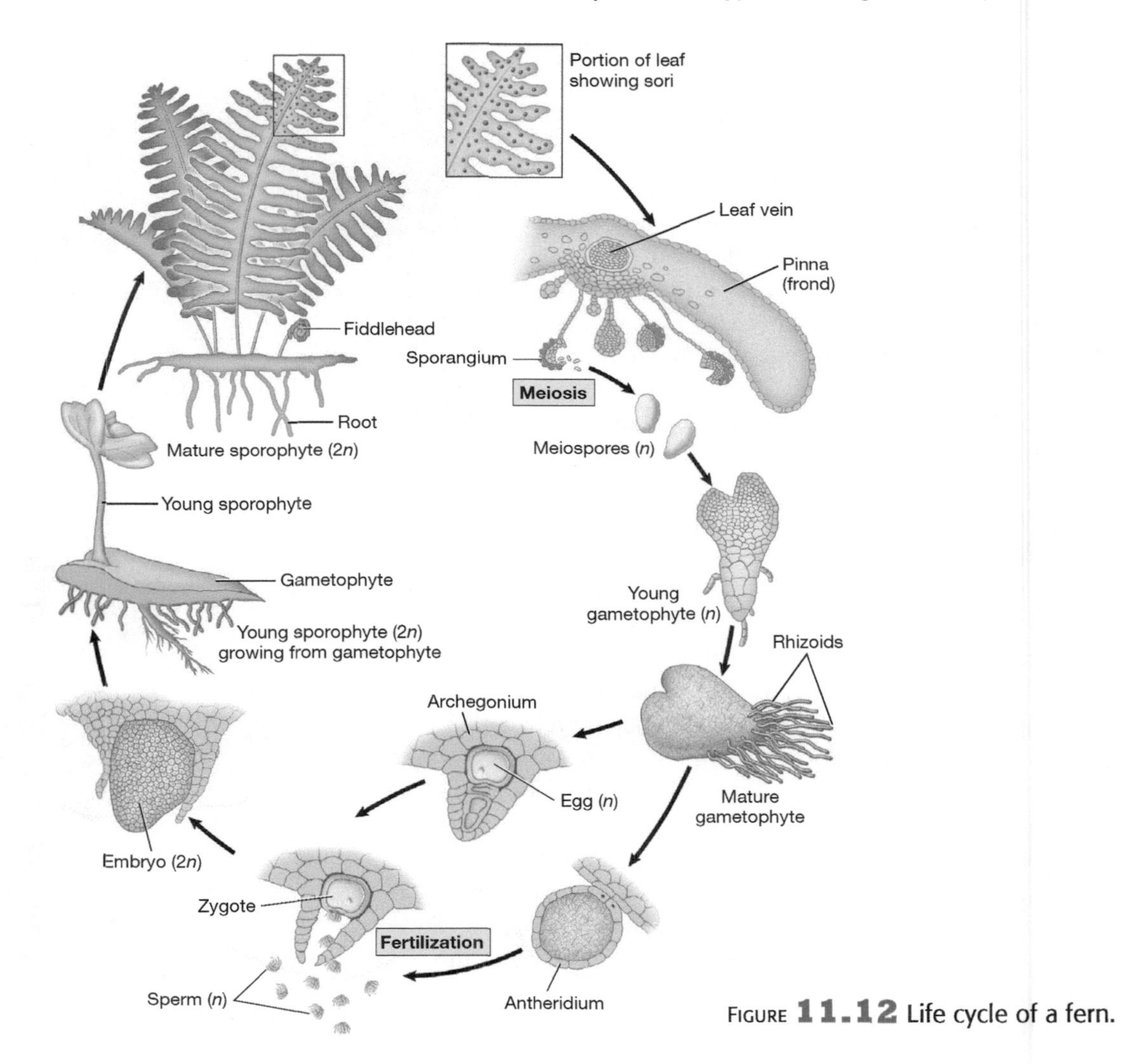


FIGURE **11.12** Life cycle of a fern.

## Ferns—Complete Drawing #3 in the Handout on Alternation of Generation in Plants after drawing these slides.

| From the live specimens, draw a **fern frond**. **Is this the sporophyte or the gametophyte**? | From the live specimens, draw t**he sori and the spores**. **What process produced the spores**? |
|---|---|
| From prepared slides, draw the **fern gametophyte**. Label the **antheridia**, **archegonia** and **rhizoids**.<br><br>**How large is this gametophyte**? | From prepared slides, draw **the fern gametophyte with an attached sporophyte**. |

**Two other groups of seedless vascular plants are frequently observed in southwest Virginia, often in the same habitats as ferns.**

**Horsetails** (Genus ***Equisetum***) once grew as large are modern trees and, along with Lycophytes, formed our coal deposits. Today's species are smaller, generally less than 4-feet tall and grow in wet, marshy habitats. Horsetails have hollow, jointed stems impregnated with silica, which gives them another common name, **scouring rushes.** In earlier times, these plants were used to scour out cast iron pots and pans.

**Examine and draw the live specimens of horsetail. Label the strobilus (the spore-producing structure, if present), the nodes, the leaves and branches.**
**Is this horsetail the sporophyte or the gametophyte?**

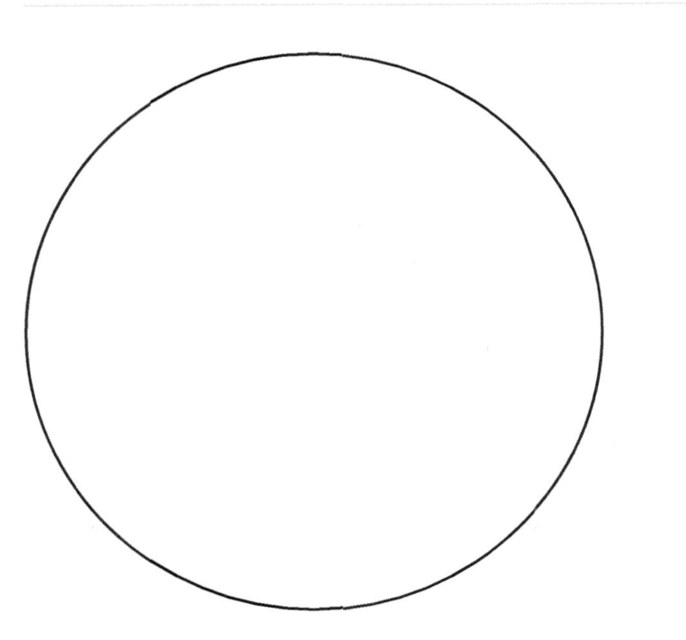

**Lycophytes** (**Genus *Lycopodium***) are small (usually less than 10 inches) evergreen plants that are common in temperate forests. Their common names include "club moss" (because of their superficial resemblance to a moss gametophyte) and "ground pine" (because some species common in pine forest may resemble small trees). But lycophytes are neither moss nor pines. Instead, they are seedless vascular plants with life cycles much like those of ferns and horsetails.
**Examine and draw the live specimens of *Lycopodium*. Label the strobilus (the spore-producing structure, if present).**
**Is this *Lycopodium* the sporophyte or the gametophyte?**

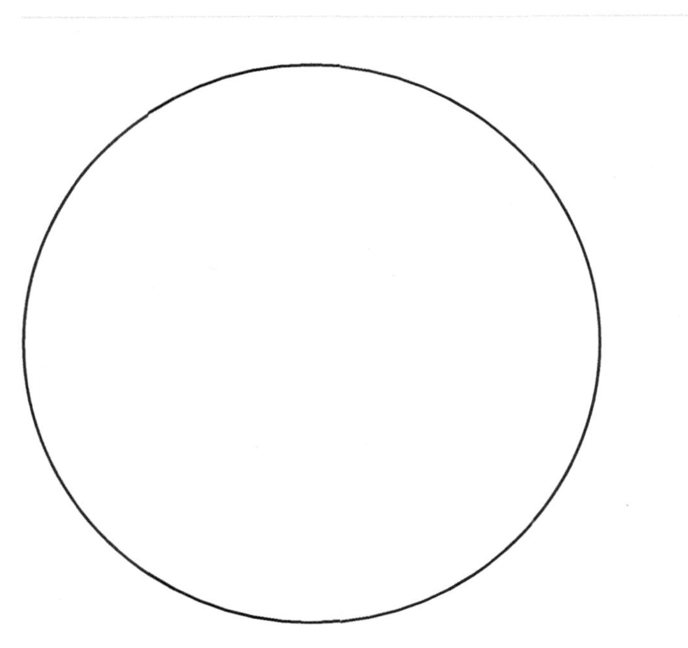

# Review Questions

**1** Let's review the differences between bryophytes and ferns by checking all the boxes that apply.

| Characteristic | Bryophytes | Ferns |
|---|---|---|
| Multicellular embryos are present. | | |
| Spores are present. | | |
| Sperm reach the egg by swimming. | | |
| The gametophyte is larger than the sporophyte. | | |
| The sporophyte is larger than the gametophyte. | | |
| Vascular tissue is present. | | |
| Cytoskeleton is present. | | |

**2** Key terms to define.

a. sporophyte ______________________________.

b. gametophyte ______________________________.

c. haploid ______________________________.

d. diploid ______________________________.

e. spore ______________________________.

f. archegonium ______________________________.

g. antheridium ______________________________.

h. fertilization ______________________________.

i. zygote ______________________________.

# The Green Machine II
## Understanding Seed Plants

# 12

*For in the true nature of things, if we rightly consider, every green tree is far more glorious than if it were made of gold and silver.*
—**Martin Luther (1483–1546)**

## OBJECTIVES

*At the completion of this chapter, the student will be able to:*

1. Compare and contrast gymnosperms and angiosperms.
2. Explain the basic biology and reproduction of phyla Ginkgophyta and Coniferophyta.
3. Distinguish springwood, summerwood, heartwood, and sapwood and determine a tree's age by annual rings.
4. Describe the organization, fundamental characteristics, and life cycles of selected gymnosperms.
5. Describe the organization, fundamental characteristics, and life cycles of selected angiosperms.
6. Compare and contrast and provide examples of basal dicots, eudicots, and monocots.

## Just wondering . . .

*Answer the following questions prior to coming to lab.*

1. Are there any old wives' tales regarding plants that are true, such as chewing willow bark will relieve a toothache?

2. Why does my hydrangea change colors?

3. What are five medically significant plants?

4. What is the main reason why sundews, pitcher plants, and other "carnivorous" plants consume insects?

5. What is catnip, and how does it work?

# Gymnosperms

Just look around—most of the plants in the world are seed plants. When you are picnicking on soft grass, taking a stroll in the park admiring the ornamental azaleas and cycads, walking through the majestic forest appreciating the splendor of the giant oaks and pines, shopping for fruits and vegetables at the local market, or just smelling the roses—seed plants are all around you. Today, the seed plants constitute the majority of plants on Earth, with an estimated 300,000 species. Seed plants first appear in the fossil record in the late Devonian period, approximately 360 million years ago (Fig. 12.1). Paleobotanists agree that the gymnosperms, particularly the conifers, flourished in the Permian period.

FIGURE **12.1** Fossil *Ginkgo biloba* leaf impression from Paleocene sediment found in Morton County, North Dakota.

The seed plants, **spermatophytes**, of today are divided into two major groups, the **gymnosperms** (naked seed leaf or cone bearing plants) and the **angiosperms** (flowering plants). The gymnosperms consist of the following four living phyla (Fig. 12.2):

1. Cycadophyta, the cycads and sago palms
2. Ginkgophyta, only one living species, *Ginkgo biloba*
3. Gnetophyta, three genera of unusual plants including *Ephedra*, *Welwitschia*, and *Gnetum*
4. Coniferophyta, the largest phylum, including pine, spruce, sequoia, juniper, cedar, and cypress

The seed plants feature a life cycle dominated by the sporophyte generation. Examples of this generation are the giant redwood and the tiny duckweed. The sporophyte produces two distinct types of gametophytes and is **heterosporous.** Multicellular male gametophytes (microspores) are called **pollen grains.** In nature, **pollination** occurs when pollen is carried from the male reproductive organs to the female reproductive organ in a number of ways, including wind, insects, and birds. *So that's how Mendel did it!*

In seed plants a pollen tube forms, allowing the sperm in the pollen grain to unite with the egg in the ovule. The ovule is a sporangium enclosed by modified leaves called the **integument.** The fertilized egg becomes the embryo, and the ovule's integument forms a protective seed coat. The **seed** provides the embryonic plant essential nourishment and protection. Thus, the seed can withstand harsh conditions and stay dormant for many years.

FIGURE **12.2** Example gymnosperms: (**A**) phylum Cycadophyta; (**B**) phylum Ginkgophyta; (**C**) phylum Gnetophyta; (**D**) phylum Coniferophyta.

Gymnosperm and angiosperm seeds are distinct (Fig. 12.3). The term *gymnosperm* literally means "naked seed." In these plants, the seeds are not enclosed in an ovule, and they mature on the surface of a cone scale, such as that of a pine cone. The nutritive material in gymnosperms accumulates prior to fertilization. In angiosperms, the nutritive material is stored only after fertilization. The seeds of angiosperms are encased in a fruit. In both cases, the parental sporophyte generation provides nutrition to potential offspring, giving them a distinct advantage over those of seedless plants.

Evolution of the seed has changed the destiny of plants as well as humans. Seeds have allowed seed plants to become the dominant plants on Earth by allowing them to literally "get a head start" on life. Seed plants provide food for animals and humans. Neolithic human societies approximately 12,000 years ago used seed plants, such as wheat, figs, corn, and squash, and shaped their destinies through artificial selection.

The gymnosperms first appear in the fossil record approximately 305 million years ago during the Carboniferous period. During the Mesozoic era, the gymnosperms dominated plant life. Although angiosperms dominate Earth today, the gymnosperms are still important plants in many ecosystems. The gymnosperms generally lack flowers and fruits.

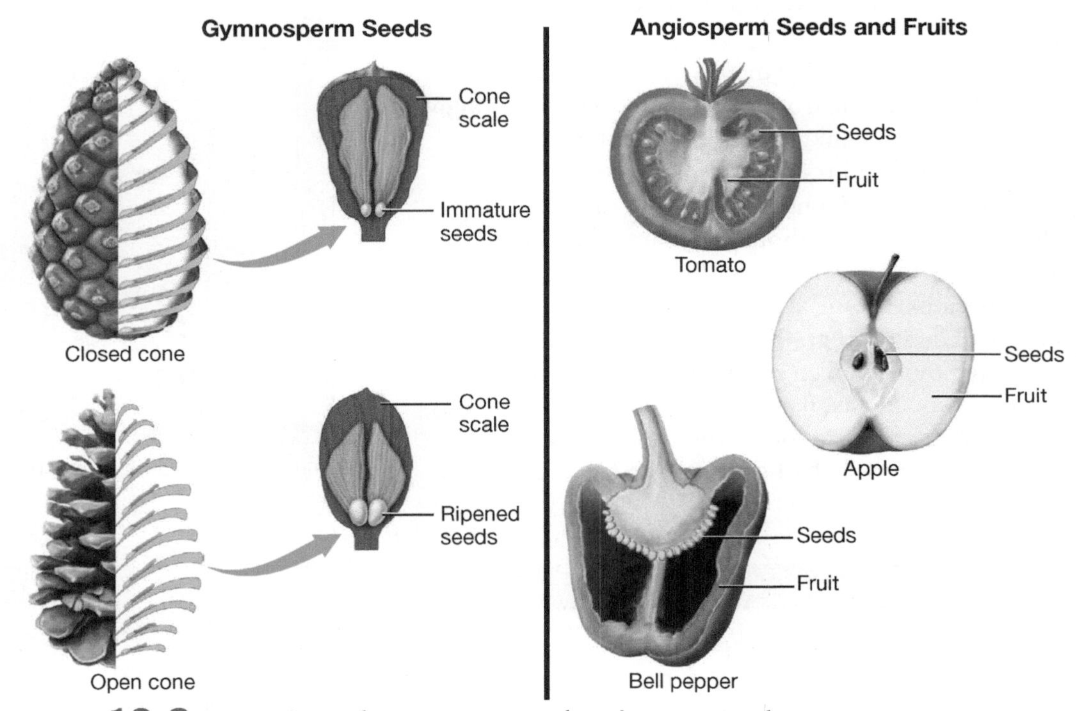


FIGURE **12.3** Comparison of gymnosperm and angiosperm seeds.

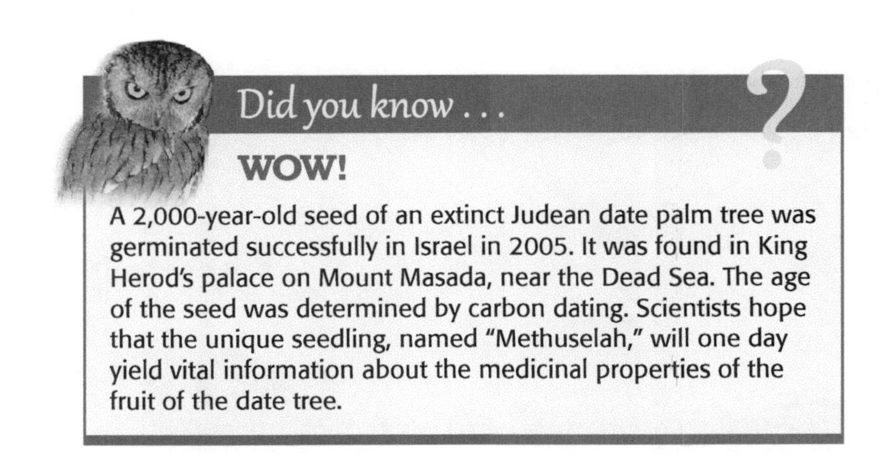


**WOW!**

A 2,000-year-old seed of an extinct Judean date palm tree was germinated successfully in Israel in 2005. It was found in King Herod's palace on Mount Masada, near the Dead Sea. The age of the seed was determined by carbon dating. Scientists hope that the unique seedling, named "Methuselah," will one day yield vital information about the medicinal properties of the fruit of the date tree.

# EXERCISE 12.1 Phylum Ginkophyta

How did you do on your last quiz? Oh no! Some people claim you should have taken *Ginkgo biloba* to improve your memory. Others claim it is a gimmick. *Ginkgo biloba* is the last living member of phylum Ginkgophyta (Fig. 12.4). The first member of the genus Ginkgo appeared in the Jurassic period, but by the Pliocene epoch of the Cenozoic era, with the exception of a small population in central China, all ginkgos had become extinct.

Charles Darwin termed *Ginkgo biloba* a living fossil. Modern *Ginkgo biloba* trees are thought to have descended from seeds collected in a Japanese temple garden. Today, *Ginkgo biloba* is known as the maidenhair tree because its distinct, notched, fan-shaped leaves resemble the pinnae of maidenhair ferns. In Chinese, the term *ginkgo* literally means "silver apricot." But the seed smells so bad that female ginkgos have been called stink-bomb trees! Most ginkgo trees planted in populated areas are male because of the females' nauseating odor. The nut within the seed, though, is tasty and is prized in Asia.

Ginkgo trees can reach a height of 30 meters or more. The trunk can exceed 3.5 meters in diameter. The trunks of ginkgo trees are straight, columnar, and sparingly branched. Ginkgo trees are popular in cities because they are beautiful, hardy, and resistant to many pests. They have an appealing growth pattern and are thought to improve air quality. *Ginkgo biloba* may live for more than a thousand years. The oldest ginkgo tree, in China, is more than 3,500 years old.

Ginkgo trees are dioecious, having separate sexes. Male trees produce pollen in their strobili. Pollen grains housing sperm cells are carried by wind to a waiting ovule. Pollen tubes form and travel through the ovule. When the pollen tube bursts, flagellated sperm make their way to the egg for fertilization. After fertilization, embryos form, and the integument develops into an extremely bad-smelling, fleshy seed coat.

FIGURE **12.4** Ginkgo, or maidenhair tree.

## Procedure 1
### Macroanatomy of *Ginkgo biloba*

*Ginkgo biloba* is a unique plant in many ways. The leaves do not possess a midrib (central vein) and have dichotomous (forked) venation (Fig. 12.5). The tree is deciduous (shedding leaves yearly), and the leaves turn bright yellow before **abscission** (shedding) in the fall. Ginkgo trees have two types of shoots: short shoots, or spurs, appear knobby and feature clusters of leaves and immature ovules. The leaves of slow-growing, short shoots usually are unlobed or slightly bilobed. The leaves of fast-growing, long shoots usually are deeply bilobed.

**Materials**
- ❑ Dissecting microscope
- ❑ Fresh or preserved stems and leaves of *Ginkgo biloba*
- ❑ Colored pencils
- ❑ Camera

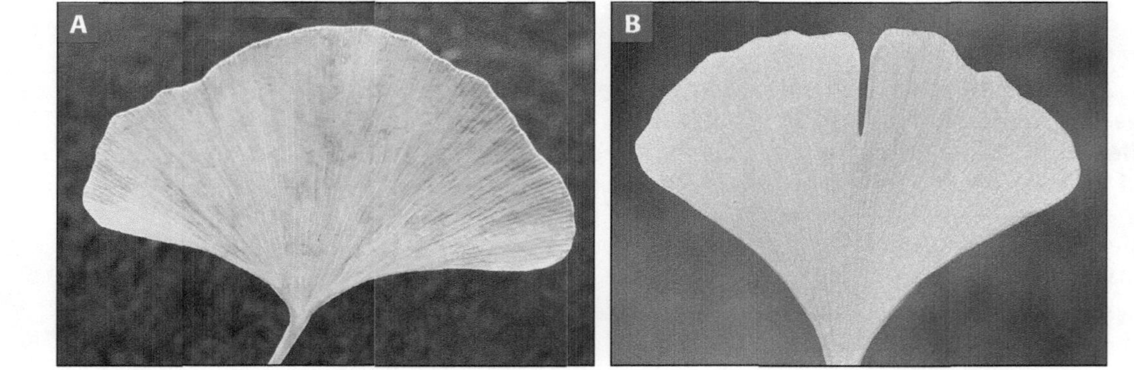


FIGURE **12.5** Leaves from the *Ginkgo biloba* tree: (A) unlobed; (B) bilobed.

**1** Procure living or preserved specimens of *Ginkgo biloba*.

**2** Using a dissecting microscope, observe the anatomical features of your specimens. Describe and sketch the short shoots, long shoots, leaves, and, if available, the sexual structures (Figs. 12.6–12.9). Place your labeled sketch and observations in the space provided below.

FIGURE **12.6** Leaves and immature ovules on a short shoot of the ginkgo tree, *Ginkgo biloba.*

FIGURE **12.7** Pollen strobili of the ginkgo tree, *Ginkgo biloba.*

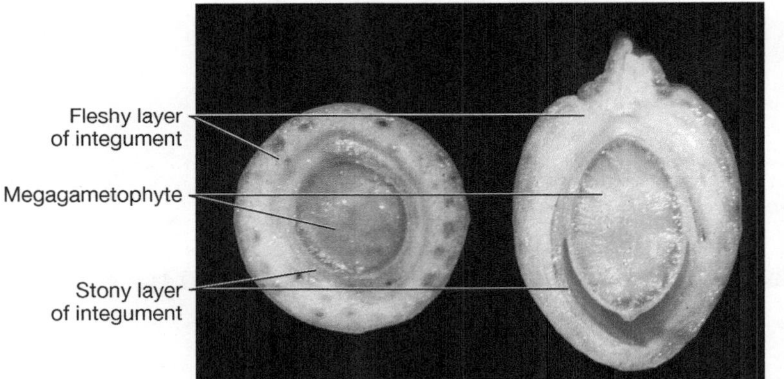


FIGURE **12.8** Transverse and longitudinal sections through a living immature seed of *Ginkgo biloba* showing the green megagametophyte.

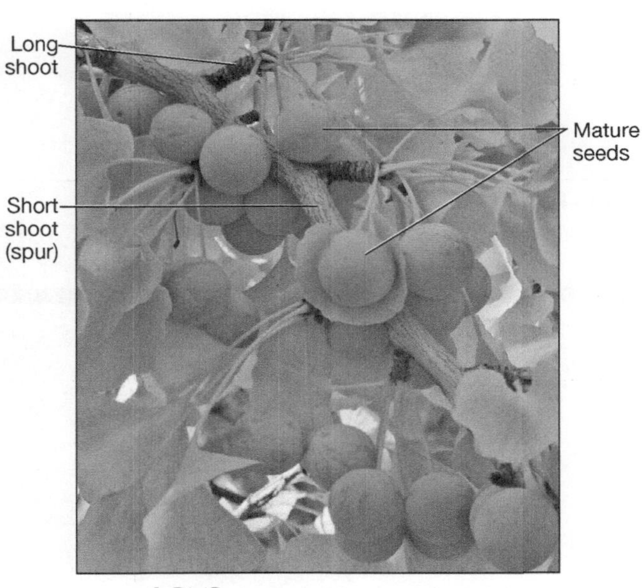


FIGURE **12.9** Branch of a *Ginkgo biloba* tree supporting a mature seed.

Shoots and leaves of *Ginkgo biloba* ______________________

______________________________________

______________________________________

# Check Your Understanding

**1.1** Why are female ginkgo trees considered undesirable in urban areas?

**1.2** Sketch and describe a ginkgo leaf.

*Ginkgo* leaf

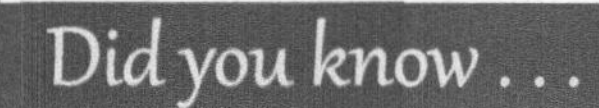


### Of Knots and Knees

Look at a piece of flooring or furniture, and observe the knots. The pattern of knots seems to give wood its character. A knot is where the base of a branch has been overtaken by the lateral growth of the trunk. The bald cypress *(Taxodium distichum)* is characterized by the presence of knees, or pneumatophores. Although the exact function of cypress knees is unknown, they are thought to provide stability in wet soils and aerate the roots, providing oxygen.

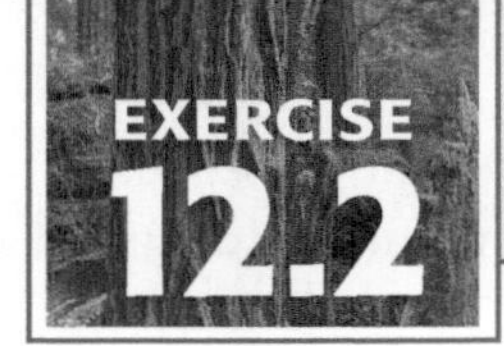

# EXERCISE 12.2 Phylum Coniferophyta

Pines, cypresses, spruces, redwoods, cedars, hemlocks, junipers, and yews are common gymnosperms placed in phylum Coniferophyta (Fig. 12.10). This phylum is composed of approximately 600 species of woody, mostly evergreen, cone-bearing plants most often found in cold and temperate climates.

Some conifers are record-setters. *Pinus longaeva*, the Great Basin bristlecone pine, is the oldest-known, nonclone, living organism on Earth. One specimen, known as "Methuselah," found in eastern California is nearly 5,000 years old. Also located in California, a coastal redwood, *Sequoia sempervirens*, is the tallest tree in the world, measuring more than 110 meters in height. The "General Sherman," *Sequoiadendron giganteum,* a giant sequoia in California, measures more than 83 meters in height and has a circumference at the ground exceeding 31 meters. It is the largest tree on Earth by volume.

Some conifer species are sources of lumber, paper, wood alcohol, turpentine, and resin. Several species, such as juniper and yew, are ornamentals. In this regard, bonsai conifer plants are popular, along with spruce and fir Christmas trees. Oils from conifers are used in soaps and air fresheners. Humans eat some seeds, such as pine nuts. Several conifer products, such as taxol (a cancer treatment), are used in medicine.

In the seed cone, the ovules consist of **megasporangia** located on the upper surface of each of the individual scales. As the result of meiosis, four megaspores are produced in each megasporangium, but only one megaspore survives (Fig. 12.11). This megaspore undergoes mitosis and eventually forms a mature female gametophyte. The female gametophyte possesses between two and six **archegonia.** Each individual

FIGURE **12.10** Various conifers: (A) bristlecone pine, *Pinus longaeva*; (B) juniper, or arbor vitae; (C) Colorado blue spruce, *Picea pungens*.

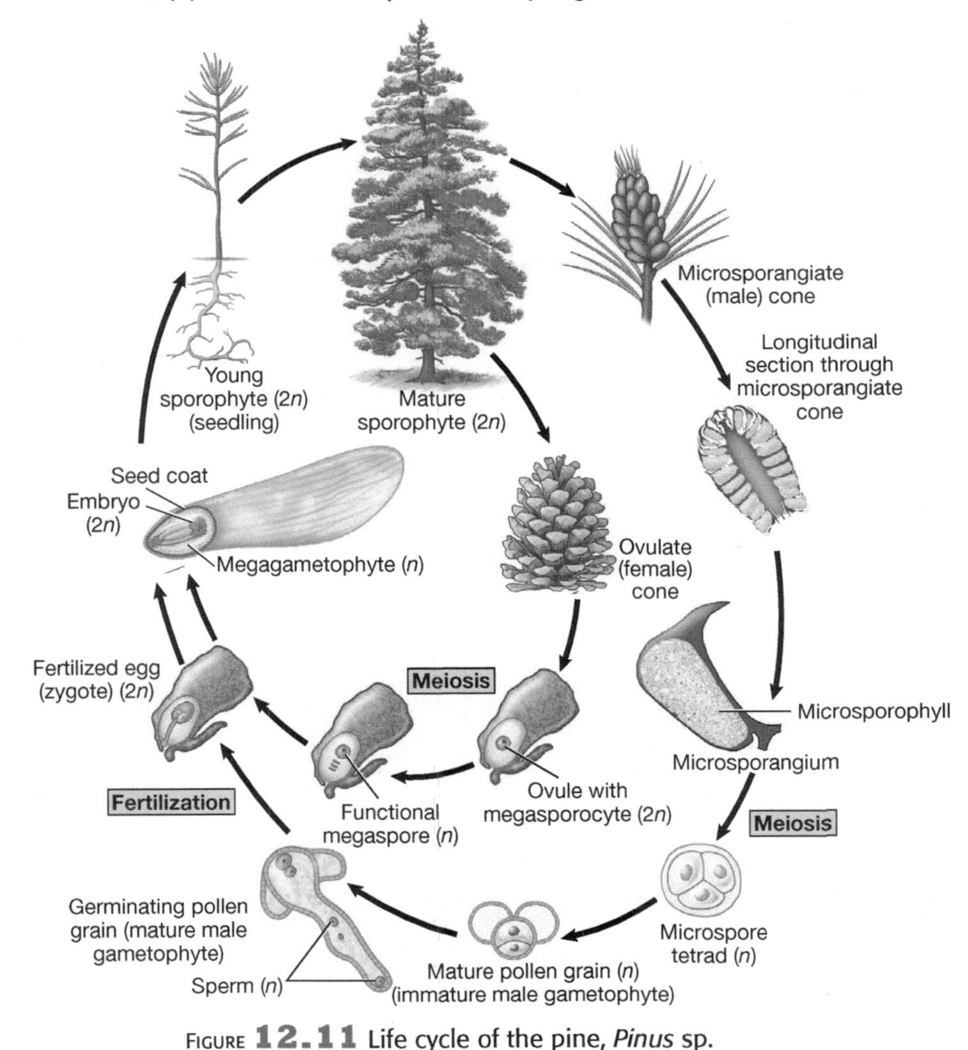


FIGURE **12.11** Life cycle of the pine, *Pinus* sp.

archegonium contains a large egg near the opening of the ovule. Initial seed cones are small, scaly, and slightly opened to allow pollen to enter the ovule-bearing scales. Pollen enters the mature seed cone and forms sperm cells ultimately delivered to the waiting egg by means of a pollen tube. The egg is fertilized and forms a zygote. The seed cone closes and increases in size as the seeds develop.

## Procedure 1

### Macroanatomy of Conifers

**Materials**

- ❑ Dissecting microscope or hand lens
- ❑ Specimens of select conifers, such as pine, bald cypress, cedar, spruce, juniper, arborvitae, fir, and other specimens provided by the instructor
- ❑ Colored pencils
- ❑ Camera

1. Procure needed equipment and supplies.
2. Observe the overall specimens, leaf arrangement, bark, cones, and distinguishing characteristics. Compare your observations with the images of conifers in this chapter. Place your observations and sketches in the space provided below.

Specimen ______________________

______________________

______________________

Specimen ______________________

______________________

______________________

Specimen ______________________

______________________

______________________

Specimen ______________________

______________________

______________________

Specimen ______________________

______________________

______________________

## Procedure 2

### Macroanatomy of Conifer Cones

Many seed cones, such as those in pine, are woody (Figs. 12.12–12.13). Others, such as those in junipers, are fleshy. The seeds of conifers released from the seed cone are winged and require air dispersal. The dry, scaly, woody cones beneath a pine tree represent the spent seed cones. After a seed lands on a suitable substrate, it germinates and develops into a new sporophyte.

**Materials**

- ❑ Dissecting microscope
- ❑ Colored pencils
- ❑ White paper
- ❑ Sterile microscope slides
- ❑ Dropper
- ❑ Water
- ❑ Coverslip
- ❑ Pollen cones (male/staminate) from select conifers
- ❑ Seed cones (female/ovulate) from select conifers
- ❑ Seeds from select conifers
- ❑ Camera

FIGURE **12.12** Images of the seed cones, or female cones, of conifers: (A) pine, *Pinus* sp.; (B) fir, *Abies* sp.; (C) spruce, *Picea* sp.; (D) bald cypress, *Taxodium* sp.; (E) yew, *Taxus* sp.; (F) arbor vitae, *Thuja* sp.

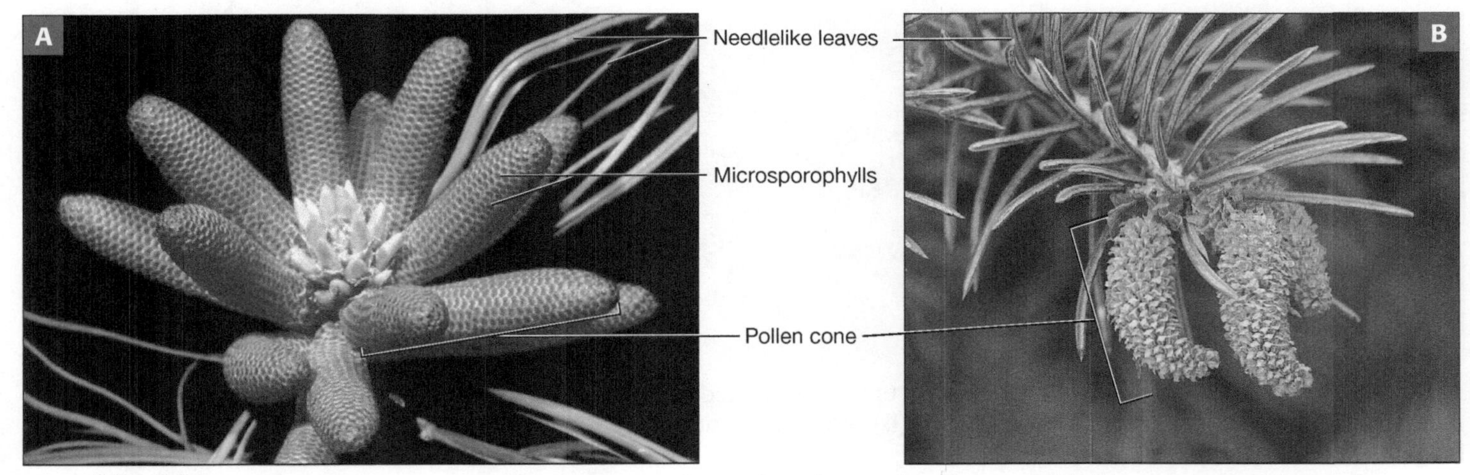


FIGURE **12.13** Microsporangiate cones of (A) *Pinus* sp. prior to the release of pollen, and (B) *Picea pungens* after pollen has been released. The pollen cones, or male cones, are at the end of a branch.

## Angiosperms

Charles Darwin (1809–1882) called the sudden appearance of modern flowers in the fossil record "an abominable mystery" because their abrupt emergence was difficult to explain. Since the time of Darwin, however, the fossil record has increased tremendously. The discovery in 1998 of an angiosperm fossil collected in China—*Archaefructus sinensis* from the Cretaceous period approximately 125 million years ago—and other discoveries have provided clues as to how angiosperms evolved. These steps began in the Jurassic period, and today the flowering plants are the dominant plants on Earth, representing more than 90% of all living plant species and 18% of all species. Figure 11.1 is a cladogram of the plants highlighting the position of the angiosperms. Development of the angiosperms transformed the face of the planet. Today the landscape is painted with colorful flowers as plants advertise themselves to pollinators. As Ralph Waldo Emerson said, "The Earth laughs with flowers."

The **angiosperms** (Greek = vessel, seed), known as the flowering plants, are the most diverse and numerous group of plants on Earth. Botanists have identified more than 280,000 species of angiosperms thus far. As in the gymnosperms, the sporophyte generation is the dominant portion of the life cycle of angiosperms. Angiosperms can be found in a number of environments. They range in size from duckweed *Wolffia angusta*, smaller than 1 millimeter in diameter, to Australia's 100-meter tall mountain ash tree, *Eucalyptus regnans*.

Angiosperms vary in form, including delicate orchids, strange insectivorous sundews, succulent cacti, and the majestic baobab tree. Most angiosperms are autotrophic. Several species of angiosperms are parasitic. Mistletoe (*Phoradendron* sp.) is a hemiparasite undergoing photosynthesis and parasitizing its host plant, and dodder (*Cuscuta* sp.) is a true parasitic plant. Indian-pipe (*Monotropa uniflora*) and snowplant (*Sarcodes sanguinea*) are two of several saprophytic species of angiosperms. Spanish moss (*Tillandsia usneoides*) and some orchids, cacti, and ferns are epiphytes, or "air plants," that attach to a substrate, such as another plant or the side of a building. The angiosperms are important to humans as sources of food, medicine, aesthetic beauty, cotton, lumber, and other products.

The angiosperms are noted for their reproductive structures called **flowers**. Flowers are the exclusive reproductive organs of angiosperms. Although we cherish their beauty, they are elaborate reproductive structures solely programmed to propagate the species. Their color, texture, and nectar are designed to ensure pollination (Fig. 12.14). Evolutionarily speaking, the appearance of flowers during the Cretaceous period changed the face of the planet and opened the door for the coevolution of many pollinators.

Most angiosperms are deciduous, losing their leaves during the winter or perhaps during a drought. Angiosperms such as peas are **herbaceous**, possessing little or no woody tissue, and others, such as an apple tree, are **woody**. Flowering plants can cycle from germination to mature plant in less than a month or as long as 150 years or more. In annuals, such as zinnias, the life cycle of a plant is completed in one season. Biennials, such as parsley, complete their life cycle in two growing seasons. Perennials, such as tulips, may take more than two seasons to complete their life cycle.

FIGURE **12.14** Flowers of many angiosperms are uniquely adapted for and rely on specific animals for pollination. Example animal pollinators: (**A**) bee, *Anthophora urbana*; (**B**) broad-tailed hummingbird (female), *Selasphorus platycercus*; (**C**) lesser long-nosed bat, *Leptonycteris yerbabuenae*.

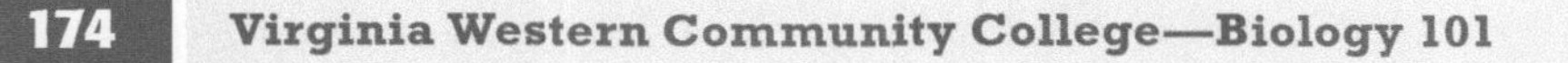

# Flowers

In 1878, Italian botanist Odoardo Beccari (1843–1920) recorded finding a corpse flower (titan arum, or *Amorphophallus titanum*) in the rainforest of Sumatra with a circumference of more than 1.5 m and a height exceeding 3 m. The fragrance of the flower is like a rotting corpse—hence its common name. The corpse flower is the largest unbranched **inflorescence** (cluster of flowers) in the world. The largest *branched* inflorescence is produced by the talipot palm, *Corypha umbraculifera,* native to Southeast Asia. Its inflorescence can attain a length of 8 m and have several million individual flowers. The largest single flower is produced by *Rafflesia arnoldii,* also known as the corpse flower. It resembles a giant, reddish-brown mushroom and can attain a diameter of more than a meter and weigh 11 kg. It is a native of the rainforest of Southeast Asia. The smallest flower in the world is produced by *Wolffia globosa,* a type of duckweed. The mature plant weighs about the same as two grains of salt. A bouquet of a dozen of these tiny flowers would be about the size of the head of a pin. Despite the tremendous range in size, color, and shape, all flowers are made primarily of the same primary components (Fig. 12.15).

FIGURE **12.15** Angiosperm flowers: (A) bird of paradise, *Strelitzia* sp.; (B) Spanish moss, *Tillandsia* sp.; (C) periwinkle, *Vinca* sp.; (D) orchard grass, *Dactylis* sp.; (E) cattail, *Typha* sp.

## Flower Anatomy

Flowers begin to develop from a specialized stalk, the **peduncle,** or from several smaller stalks, **pedicels.** The peduncle or pedicels form the **receptacle,** a swollen region that contains the other floral parts arranged in **whorls.** Keep in mind that in eudicots, the flower parts are arranged in multiples of four or five, and in monocots in multiples of three. The outermost whorl is composed of leaflike, green **sepals.** The sepals, in turn, form the **calyx.** In many species the calyx serves to protect the flower while it develops within the bud. The **petals,** the most conspicuous parts of a flower, range in color, shape, size, and fragrance. The petals collectively form the **corolla.** The color of the petals and the shape of the corolla are significant in pollination. The calyx and corolla make up the **perianth.**

The "male" portion of the flower, the **stamen,** consists of a slender stalk, the **filament,** and the saclike **anther** where pollen is produced. Collectively, the stamens make up the **androecium.** Usually, anthers release pollen by splitting open (as in a daisy), but in some species (such as azaleas) the pollen is released from pores at the tip of the anther.

The most obvious female portion of the plant is the centrally located **pistil.** (Some texts refer to the pistil as a **carpel.**) It is composed of a sticky knob that receives pollen and is known as the **stigma,** which sits atop a slender tube called the **style,** ultimately leading to the **ovary** (Fig. 12.16). The position of the ovary is significant in plant classification.

An ovary eventually forms the fruit. One or more carpels serve as the main portion of the ovary. The carpel or carpels are known collectively as the **gynoecium.** Ovules, produced on the carpels, contain the female gametophyte. Generally, the number of carpels is related to the number of divisions of the stigma. For example, each section of a tomato or a grapefruit represents a carpel.

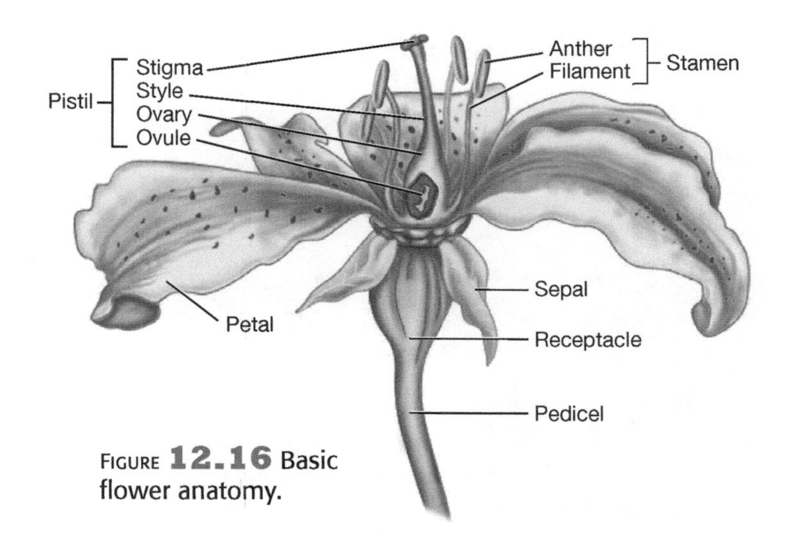


FIGURE **12.16** Basic flower anatomy.

## Flower Classification

Flower terminology and arrangement are complex. **Complete flowers**, such as magnolias, tulips, apples, azaleas, and lilies, possess sepals, petals, stamens, and pistils. **Incomplete flowers** lack one or more sepals, petals, stamens, or pistils; examples are squash, begonia, oak, and walnut (Fig. 12.17). A **perfect flower**, such as a dandelion, lily, banana, or pea, possesses both stamens and pistils. An **imperfect flower** possesses only one sex because it lacks either the stamens or the pistil (Fig. 12.18). **Staminate flowers** have only stamens, and **pistillate flowers** have only pistils. Imperfect flowers may appear on separate plants, as in holly, mulberry, and persimmon, or on the same plant, as in cattail, oak, and corn.

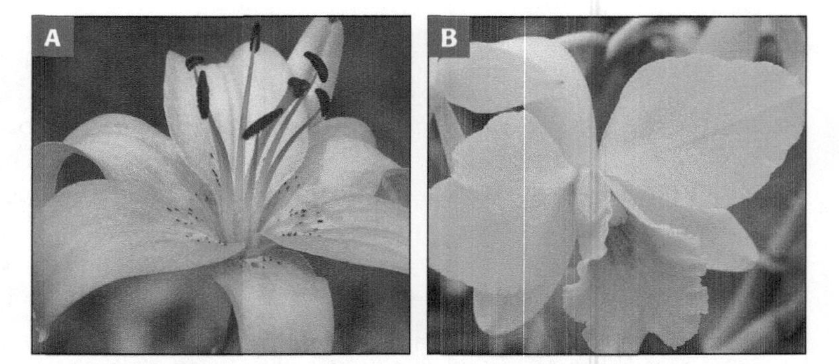

FIGURE **12.17** (A) Complete flower, lily, *Lilium* sp.; (B) incomplete flower, orchid, *Cymbidium* sp.

Floral symmetry refers to the arrangement of flowers along a plane. **Actinomorphic** (radially symmetrical) flowers can be divided into symmetrical halves by more than one longitudinal plane passing through the axis. In these flowers the petals are similar in shape and size. Examples of actinomorphic flowers are azaleas, buttercups, and roses. **Zygomorphic** (bilaterally symmetrical) flowers can be divided by a single plane into two mirror-image halves (Fig. 12.19). Zygomorphic flowers generally have petals of two or more different shapes and sizes. Examples of zygomorphic flowers are the orchid, foxglove, and snapdragon.

FIGURE **12.18** (A) Perfect flower, daisy, *Gerbera* sp.; (B) imperfect flower, pitcher plant, *Sarracenia* sp.

Flowers may be solitary, as in a petunia or camellia, or appear in clusters known as an inflorescence, as in oak, sunflower, and willow (Fig. 12.20). Catkins (Dutch = kitten) are a drooping, slim inflorescence, lacking petals or having inconspicuous petals that resemble a kitten's tail. They contain many, usually unisex, flowers arranged closely along a central stem. In some plants, such as willow, mulberry, and oak, only the male flowers form catkins and the female flowers are solitary. In other plants, such as poplar, both male and female flowers are borne in catkins.

FIGURE **12.19** (A) Actinomorphic symmetry, daffodil, *Narcissus* sp.; (B) zygomorphic symmetry, *Nemesia* sp.

FIGURE **12.20** (A) Solitary flower, dahlia, *Dahlia* sp.; and inflorescent flowers: (B) sunflower, *Helianthus* sp.; (C) walnut catkin, *Juglans* sp.

## Grasses

Many people do not realize grasses produce flowers. In the summer, just let your yard get out of control and notice the flowers and seed heads (for example, my yard while I am writing this manual!). The flowers of grasses are inflorescence. In bluegrass, wheat, rice, and other herbaceous grasses, each leaf consists of a **basal sheath** that encompasses the **culm** (grass stem) down to its point of origin, the **node**. The

**internodes** of grasses typically are hollow, such as in bamboo. The leaf blade usually grows away from the culm. A membranous scale called the ligule can be found at the junction of the basal sheath and leaf blade. The tiny projections near the base of the leaf blade are known as auricles.

A **spikelet** is a conspicuous extension of the peduncle, consisting of many small florets (Fig. 12.21). A glume, designated as the first and second glume, or protective husk, appears externally in the **floret**. A floret consists of two bracts—the lemma and the palea. Within the floret are one pistil, three stamens, the ovary, and the scalelike lodicule. After fertilization, the ovary develops into a one-seeded fruit, a grain, or caryopsis. When weeding the yard, if you cut the peduncle, the seed head will not repair.

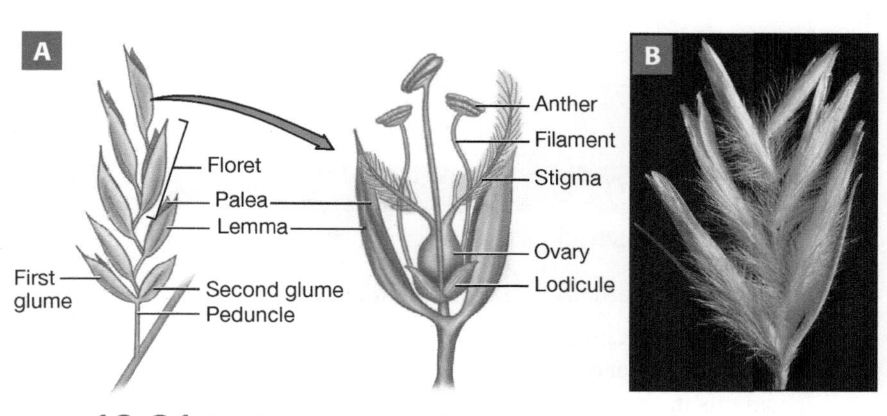


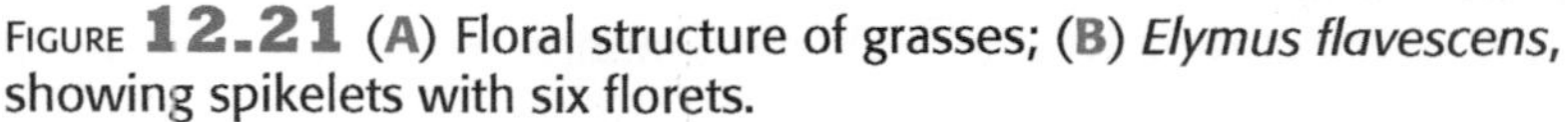
FIGURE **12.21** (**A**) Floral structure of grasses; (**B**) *Elymus flavescens*, showing spikelets with six florets.

## Pollination

The sporophyte is the dominant generation in angiosperms (Fig. 12.22). In the male portion of the plant, anthers have four pollen sacs containing numerous microsporocytes. Each microsporocyte produces four haploid microspores. After cell division, the haploid nuclei of the microspores produce a pollen grain. In the female portion of the plant, one or several ovules can be found in the ovary. Within an ovule, four megaspores are formed. One of the megaspores becomes an **embryo sac,** or female gametophyte.

During pollination, a two-celled pollen grain lands on the stigma of the same species of plant; one cell forms a tube cell and the other a generative cell (Fig. 12.23). The tube cell will form the pollen tube, and the generative cell will produce two sperm cells. The pollen tube moves down the style to the ovary with an awaiting ovule. Of the two sperm cells, one fertilizes the egg and the other fuses with two polar nuclei to form a $3n$ endosperm. This is called **double fertilization.** The endosperm nourishes the developing embryo. The ovule develops into a seed containing the embryonic plant (sporophyte) and the endosperm.

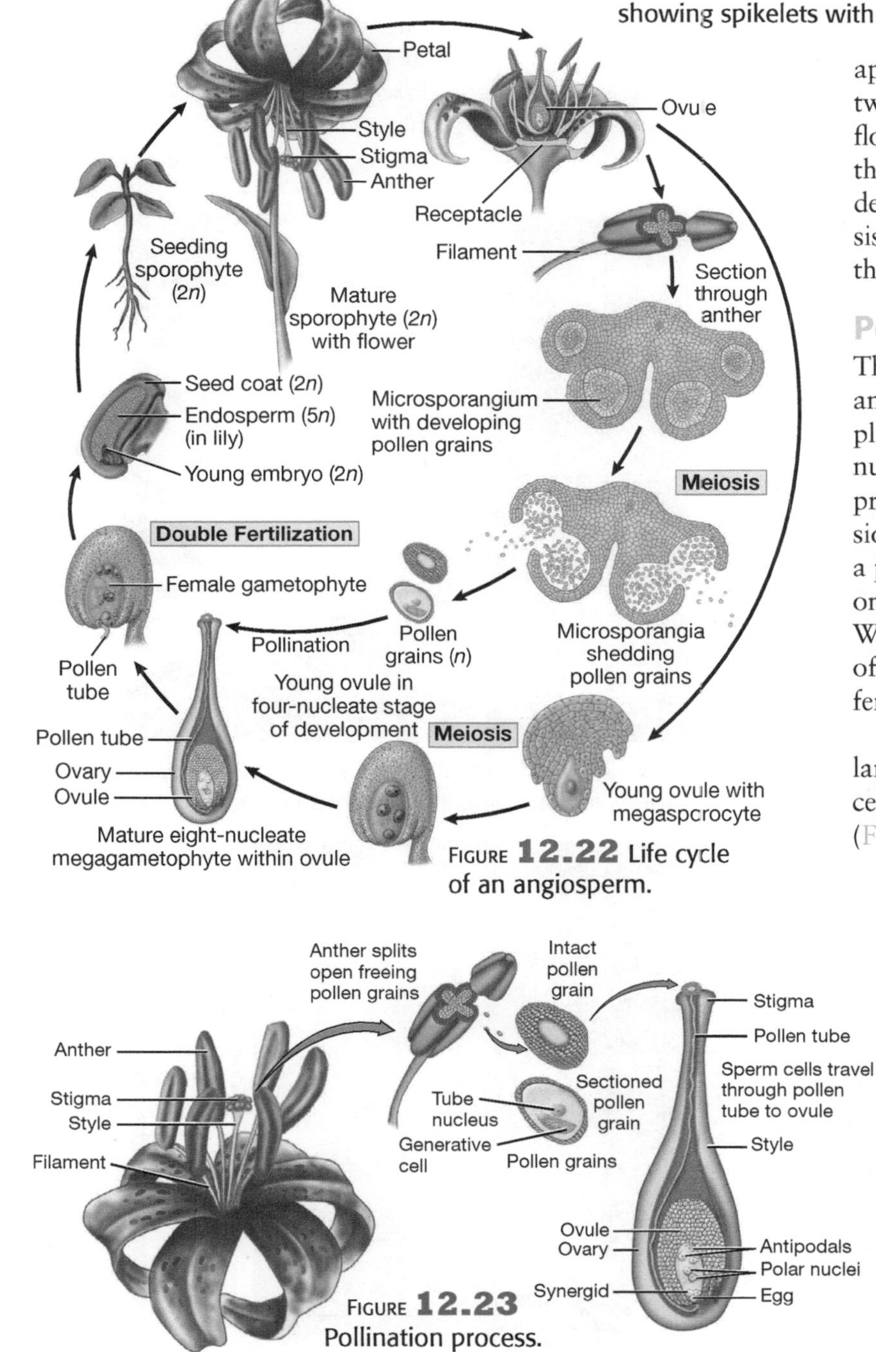


FIGURE **12.22** Life cycle of an angiosperm.

FIGURE **12.23** Pollination process.

## Procedure 1
### Flower Dissection

1 Procure the specimens and equipment.

2 Describe the specimens in detail and, using a dissecting microscope or a hand lens, observe, draw, and label the anatomical features of your specimens in the space provided on the following page (refer to Figs. 12.24–12.27). Be sure to include the common name and the scientific name of the plant in your description. Also include whether the flower is a basal dicot, eudicot, or monocot. In addition, include whether the flower is complete or incomplete, perfect or imperfect, solitary or an inflorescence, or actinomorphic or zygomorphic.

**Materials**
- ❑ Dissecting microscope
- ❑ Compound microscope
- ❑ Hand lens
- ❑ Scalpel
- ❑ Dissecting tray
- ❑ Blank slides and coverslips
- ❑ Pipette
- ❑ Flowers provided by the instructor
- ❑ Colored pencils
- ❑ Camera or camera phone (optional)



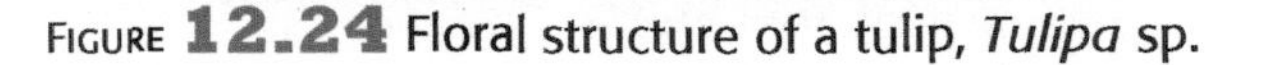

FIGURE 12.24 Floral structure of a tulip, *Tulipa* sp.

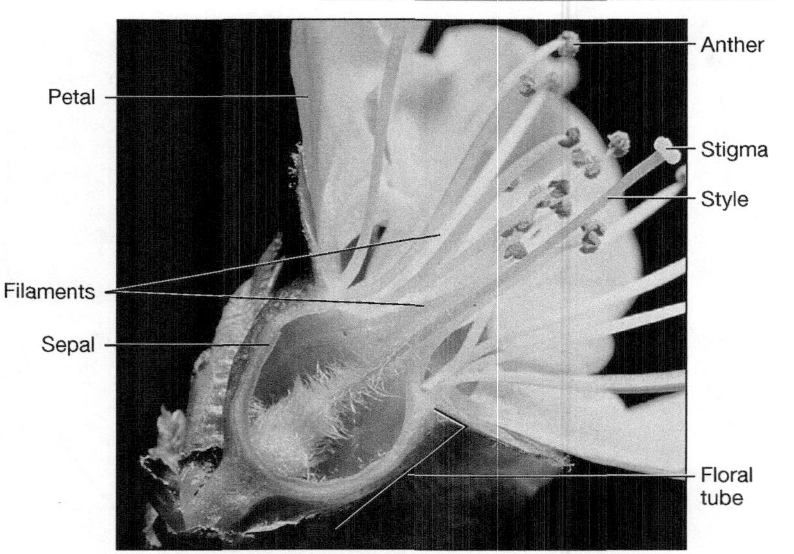


FIGURE 12.25 Structure of a dissected cherry, *Prunus* sp., showing a perigynous flower.

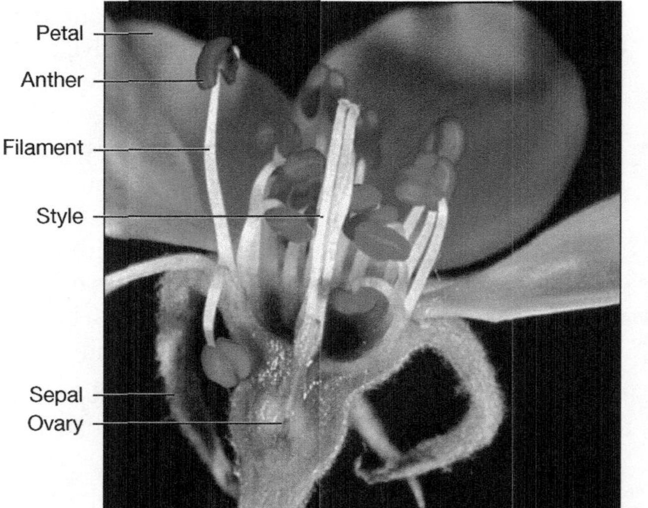
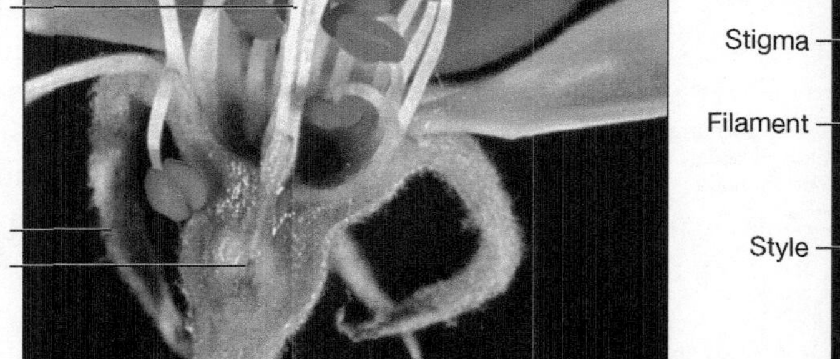


FIGURE 12.26 Structure of a dissected pear, *Pyrus* sp., showing an epigynous flower.

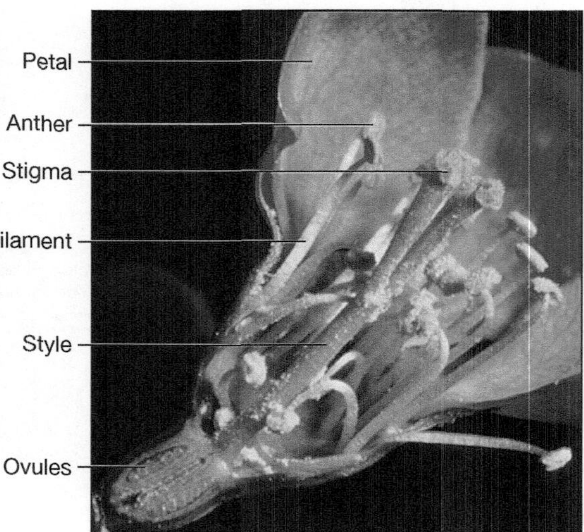


FIGURE 12.27 Dissected quince, *Chaenomeles japonica*, showing an hypogynous flower.

Specimen

Specimen

Specimen

**3** Describe the smell of your flower.

**4** Carefully remove the sepals and petals from a flower designated by the instructor. Closely observe the male and female reproductive parts, using a dissecting microscope. Carefully cut the base of the pistil and ovary longitudinally, and record your observations and sketch in the space provided (refer to Fig. 12.16).

**5** Using the point of the scalpel, remove some pollen from the anther of a flower. Make a wet mount of the pollen, and observe it under the compound microscope (Fig. 12.28). Record and illustrate your observations in the space provided on the following page.

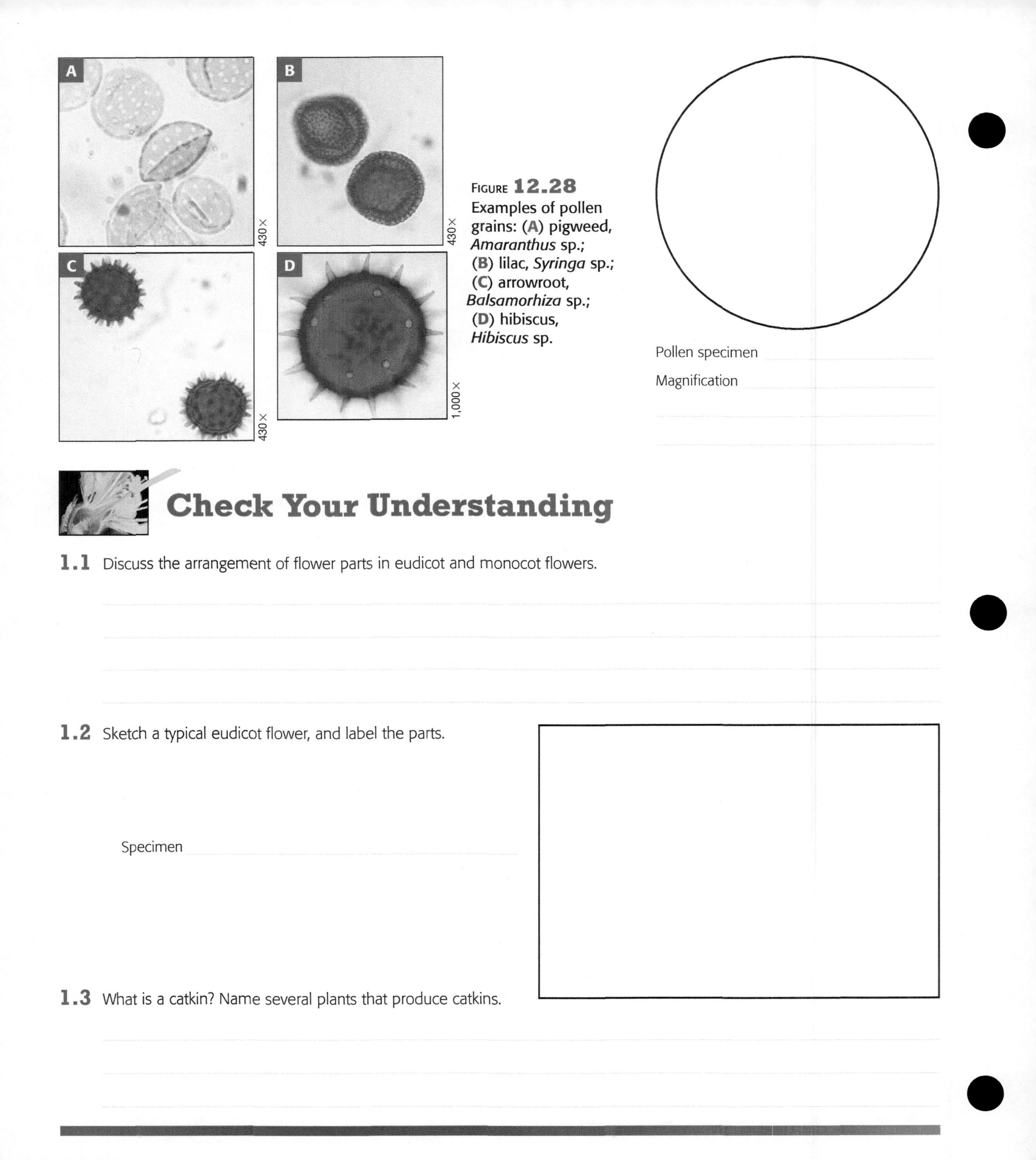


FIGURE **12.28** Examples of pollen grains: (**A**) pigweed, *Amaranthus* sp.; (**B**) lilac, *Syringa* sp.; (**C**) arrowroot, *Balsamorhiza* sp.; (**D**) hibiscus, *Hibiscus* sp.

# Check Your Understanding

**1.1** Discuss the arrangement of flower parts in eudicot and monocot flowers.

**1.2** Sketch a typical eudicot flower, and label the parts.

Specimen

**1.3** What is a catkin? Name several plants that produce catkins.

# Fruits

**Fruits** are exclusive to angiosperms. All fruits are derivatives of the ovary or ovaries of a flower and associated structures such as the receptacle. The diversity of fruits is astonishing, ranging from acorns to zucchini. Some fruits, such as the cultivated banana, seedless cucumbers, seedless watermelons, and most seedless citrus, are **parthenocarpic;** they do not require fertilization to form a fruit. In **stenospermocarpy**, both pollination and fertilization occur but the developing embryo in the seed is aborted and the seed stops developing. Seedless grapes are an example of stenospermocarpy.

Did you know . . .

### What Is a Mexican Jumping Bean?

A Mexican jumping bean is not a bean at all! It is a carpel of a seed capsule from the Mexican shrub *Sebastiana pavoniana* that houses the larva of a small gray moth called the jumping bean moth (*Laspeyresia saltitans*). While eating the nutritive material within the carpel, the larva wiggles, causing the jumping movements of the "bean."

## Fruit Anatomy

Upon maturation, the ovary of a fleshy fruit usually has three regions—the **exocarp**, the **mesocarp**, and the **endocarp**. Because these regions may merge, it sometimes is difficult to distinguish between the regions. The three regions are known collectively as the **pericarp**. In dry fruits, the pericarp may be thin, as in the hull of a peanut.

1. The exocarp, which forms the skin, or peel, of a fruit, is variable in color and texture. With its associated glands, it is called a flavedo in a citrus fruit. As an orange ripens, the outside of the flavedo changes from green (chlorophyll) to orange (mostly xanthophyll) in color.
2. The mesocarp is the fleshy portion of the fruit between the exocarp and the endocarp. In citrus fruits, the whitish region just beneath the exocarp is actually the mesocarp, called the albedo.
3. The endocarp is the inside layer of the pericarp directly surrounding the seed. The endocarp may be papery as in apples, or stony as in a peach, or slimy as in a tomato, or a shell as in a pecan. In citrus fruits the endocarp is divided into distinct segments. **Juice vesicles,** the membranous pulp of citrus fruits, provides the treasured juice. There seems to be a correlation between the number of juice vesicles and the number of segments. Thus a grapefruit will have more vesicles than a kumquat.

Did you know . . .

### How to Grow a Pineapple

A green thumb isn't necessary to grow a pineapple—just patience. This can be done by following these steps:

1. Obtain a fresh pineapple with healthy green leaves.
2. Remove several of the lower leaves to expose the stem. Cut off the crown about 3 inches below the stem. Trim any tissue from around the rim.
3. Place the crown upside down in a cool, dry, insect-free place for one week.
4. Plant the crown in an 8-inch clay pot filled with light garden soil with a 30% blend of organic matter. Be sure to form the soil around the crown up to the base of the stem.
5. Place the plant in a humid, sunny environment. Lightly water the plant weekly. (Pineapples don't like to get wet!)
6. During the summer, lightly fertilize the pineapple monthly.
7. After 18–24 months, inspect the plant's development. The pineapple will produce a red cone surrounded by blue flowers. The flowers will drop, and the fruit will begin to develop.
8. If desired, force the fruit to develop by covering the entire plant with a clear polyethylene bag and placing two ripe apples in the pot. Ethylene gas produced by the apples encourages fruit development.
9. When the pineapple matures, photograph it, and then harvest it. *Bon appetit!*

## Fruit Classification

Fruit classification is based on several features, including whether the fruit is simple or compound, fleshy or dry, and whether other floral parts are present. This scheme is not exact, and arguments abound. In any case, Table 12.1 may be helpful in classifying fruits.

TABLE **12.1** Fruit Classification

| Fruit Group | Type | Description | Examples |
|---|---|---|---|
| | | **Simple Fruits** | |
| **Fleshy fruits** | berry | ovary compound, skin from exocarp, fleshy pericarp | tomato, grape, guava, kiwi, persimmon, papaya, pomegranate, avocado. *Note:* bananas, cranberries, and blueberries are false berries. |
| | pome | accessory fruit, derived from several carpels, ovary (core) surrounded by fleshy receptacle tissue | apple, pear, quince |
| | hip | accessory fruit, derived from several carpels, encloses achenes | rose |
| | pepo | accessory fruit, berry with hard, thick rind, receptacle partially or completely encloses the ovary | squash, watermelon, cantaloupe, gourd |
| | drupe | derived from a single carpel, possesses one seed, endocarp a stony pit, exocarp a thin skin | peach, cherry, plum, olive, mango, pecan, walnut, coconut, almond, pistachio, cashew, macadamia |
| | hesperidium | berry with a leathery rind and juice sacs | orange, lemon, lime, grapefruit, kumquat |
| | | **Dry Fruits** | |
| **Dehiscent** | legume | single carpel, pod splits along two sides | peas, mimosa, bean, peanut, wisteria, redbud |
| | follicle | single carpel, splits along one side | milkweed, oleander, columbine |
| | silique | two carpels that separate at maturity, leaving a permanent partition between them | radish, mustard, cabbage, turnip |
| | capsule | composed of several carpels, separates in several ways | okra, poppy, iris, yucca, cotton, orchid, sweet gum, agave, Mexican jumping bean, Brazil nut |
| **Indehiscent** | achene | simple ovary with pericarp is dry and free from the internal seed, except at the placental attachment | sunflower, dandelion, buttercup, sycamore, buckwheat |
| | samara | simple ovary with a winged pericarp, produced in clusters | maple, ash, elm |
| | caryopsis (grain) | simple ovary, one seed with the pericarp fused to the seed coat | corn kernel, rice, oats, barley, wheat, Johnson grass, Bermuda grass |
| | schizocarp | two or more sections break apart at maturity, each with one seed | carrot, fennel, celery, dill, puncture vine |
| | nut | single seed with hard pericarp surrounded by bracts and/or a receptacle | oak (acorn), hickory, hazelnut, beech, chestnut |
| | | **Compound Fruits** | |
| **Aggregate Fruits** | | | |
| **Fleshy Fruits** | achenes | consist of a number of matured ovaries from a single flower, arranged over the surface of a single receptacle; individual ovaries are called fruitlets | strawberry, buttercup |
| | drupes | | dewberry, blackberry, raspberry, boysenberry |
| **Dry Fruits** | follicles | | southern magnolia |
| | samaras | | tulip (yellow) poplar |
| **Multiple Fruits** | | | |
| **Fleshy Fruits** | achenes | collection of fruits produced by the grouping of many flowers crowded together in a single inflorescence, typically surrounding a fleshy stem axis | fig |
| | drupes | | breadfruit, mulberry, Osage orange |
| | fused berries | | pineapple |
| **Dry Fruits** | achenes | | sycamore |
| | capsules | | sweet gum |
| | caryopsis | | corn cob and kernels |

FIGURE **12.29** Examples of simple fruits: (**A**) peach; (**B**) grapes; (**C**) apple; (**D**) pea.

**Simple fruits,** such as grapes, beans, and hickory, are derivatives of a single ovary (Fig.12.29). Many simple fruits, such as apples, oranges, and watermelons, are classified as **fleshy fruits.** Others are classified as **dry fruits** (Fig. 12.30).

In dehiscent dry fruits, the pericarp is dry, and the fruit splits at maturity; these include peas, radishes, milkweed, and orchids. Indehiscent dry fruits do not split at maturity; examples are acorns, corn kernels, parsley, and rice.

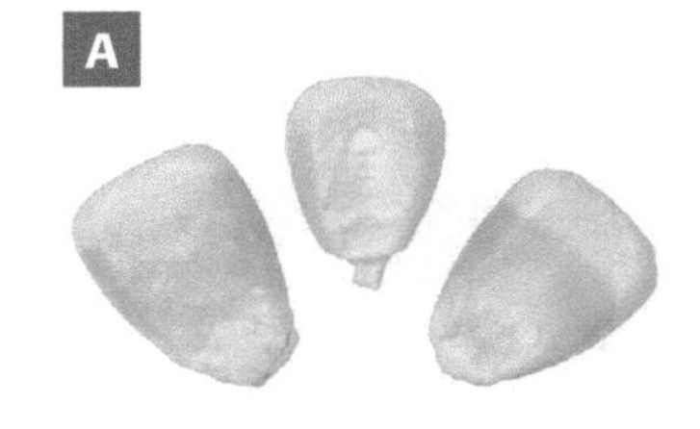

FIGURE **12.30** Examples of dry fruits: (**A**) corn; (**B**) oats.

**Compound fruits,** such as strawberries, blackberries, and figs, develop from several individual ovaries. **Aggregate compound fruits** are derived from a single flower with many pistils. In an aggregate fruit, the tiny fruitlets can be an achene or a drupe existing on a single receptacle. A strawberry is a fleshy fruit with achenes on the surface of the receptacle (the body). Look at a blackberry: each tiny drupe came from an individual ovary. Strawberries and blackberries are **aggregate fleshy fruits** (Fig. 12.31). Aggregate fruits can also be dry, such as the fruits of magnolia and yellow poplar. Multiple fruits, such as figs, pineapples, and mulberries, are derived from several individual flowers in an infloresence (Fig. 12.32).

Fruits that develop from tissues surrounded by the ovary are called **accessory fruits.** These generally develop from flowers with inferior ovaries, and the receptacle becomes a part of the fruit. Accessory fruits can be simple, aggregate, or multiple.

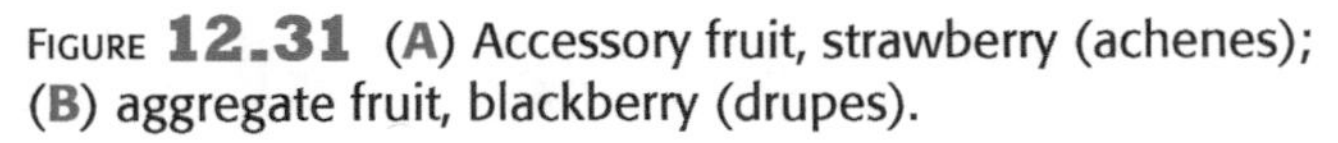

FIGURE **12.31** (**A**) Accessory fruit, strawberry (achenes); (**B**) aggregate fruit, blackberry (drupes).

FIGURE **12.32** Examples of multiple fruits: (**A**) pineapple; (**B**) fig.

EXERCISE 12.4

# Fruits

## Procedure 1

### Fruit Dissection

**Materials**

- ❑ Dissecting microscope or hand lens
- ❑ Dissecting tray
- ❑ Dissecting needle
- ❑ Forceps
- ❑ Scalpel
- ❑ Fruits provided by the instructor, such as an apple, an orange, a peach, a bean, a strawberry
- ❑ Colored pencils
- ❑ Camera or camera phone (optional)

1. Procure the specimens and equipment.
2. Describe the specimens in detail and, using a dissecting microscope or a hand lens, observe, draw, and label the anatomical features of your specimens in the space provided below. Be sure to include with your description both the common name and the scientific name of the plant as well as the type of fruit and what you infer about seed dispersal.

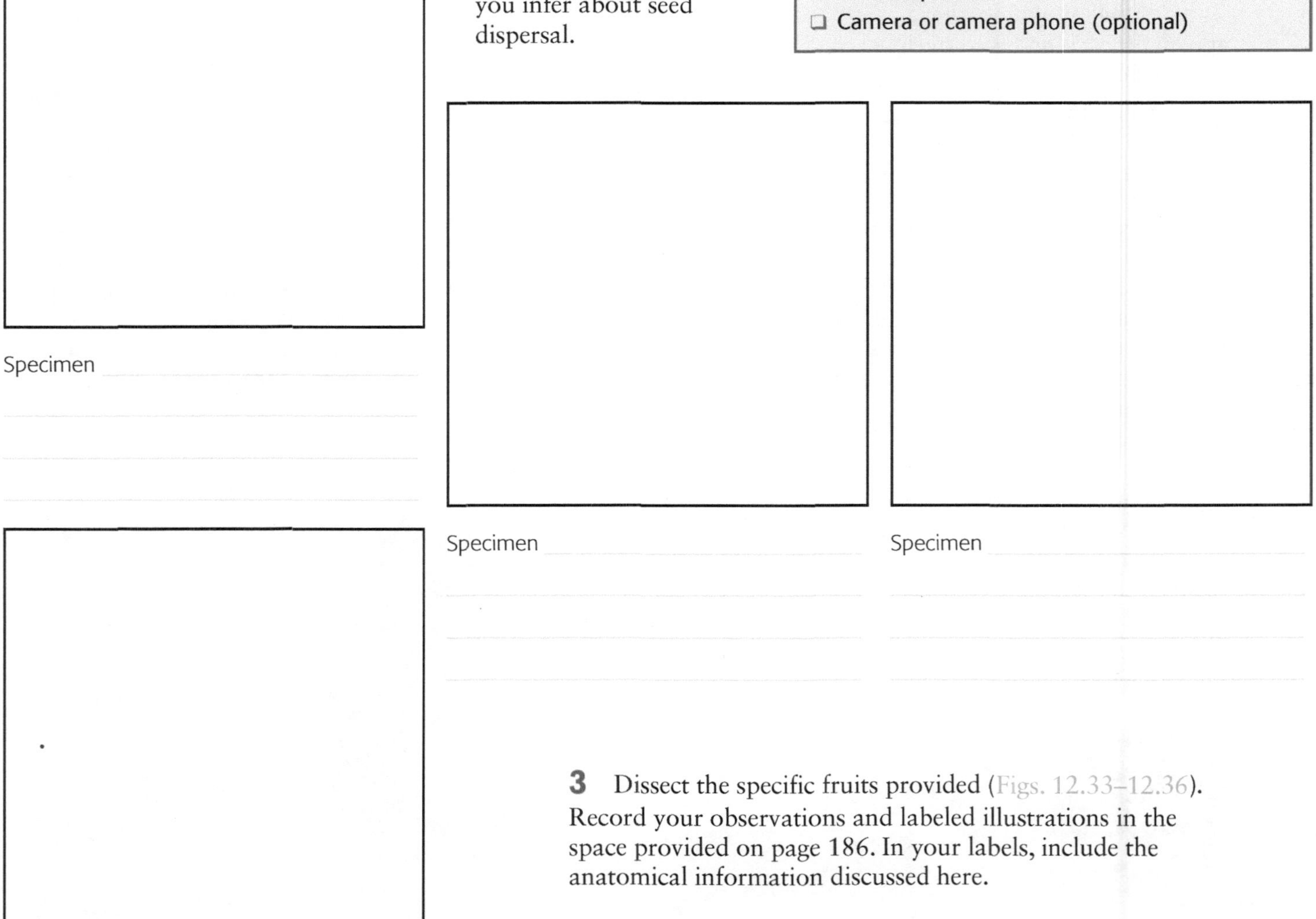


3. Dissect the specific fruits provided (Figs. 12.33–12.36). Record your observations and labeled illustrations in the space provided on page 186. In your labels, include the anatomical information discussed here.

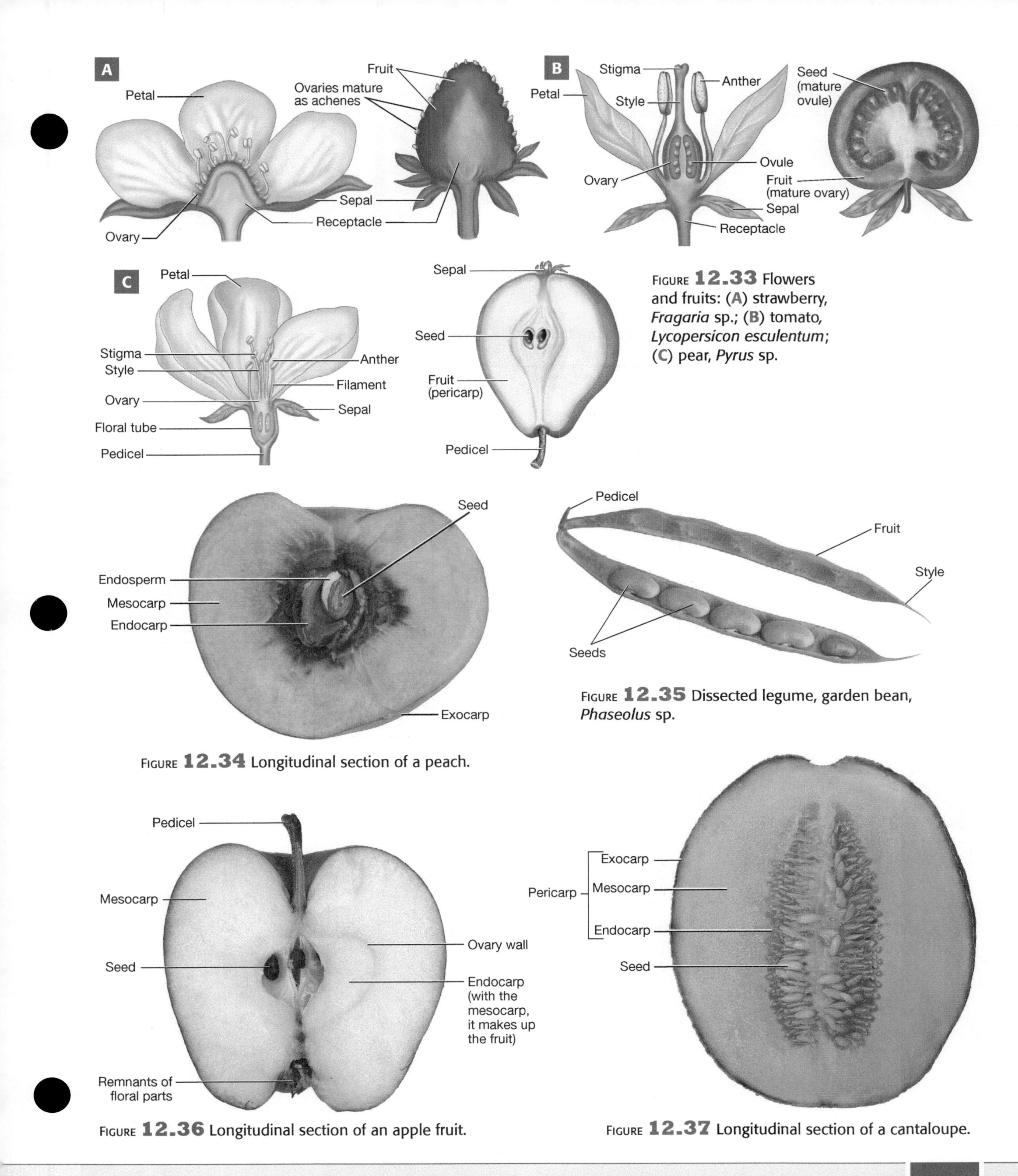


FIGURE **12.33** Flowers and fruits: (**A**) strawberry, *Fragaria* sp.; (**B**) tomato, *Lycopersicon esculentum*; (**C**) pear, *Pyrus* sp.



FIGURE **12.34** Longitudinal section of a peach.



FIGURE **12.35** Dissected legume, garden bean, *Phaseolus* sp.



FIGURE **12.36** Longitudinal section of an apple fruit.



FIGURE **12.37** Longitudinal section of a cantaloupe.

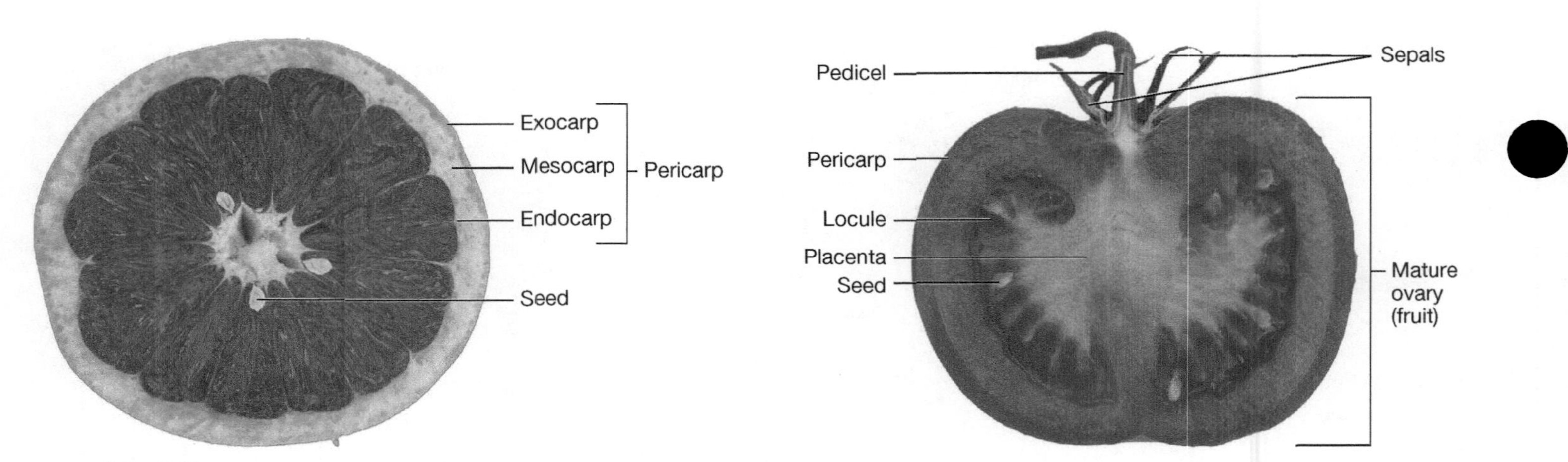


FIGURE **12.38** Transverse section through a grapefruit fruit.

FIGURE **12.39** Longitudinal section of a tomato fruit (berry).

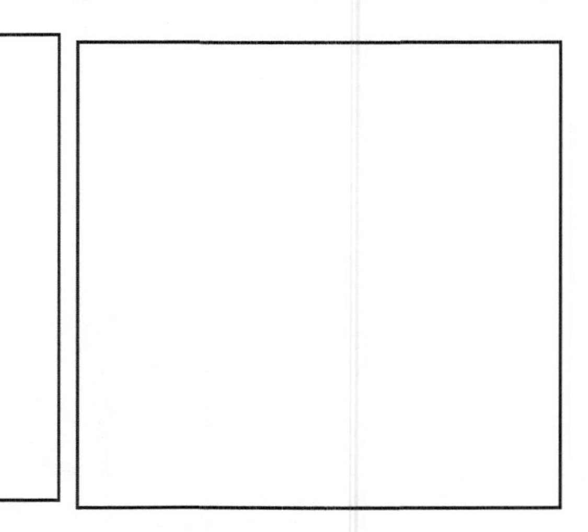

Specimen ______________

Specimen ______________

Specimen ______________

## Check Your Understanding

**2.1** What is the function of a fruit?

**2.2** Classify several fruits you've seen while browsing in a market, other than those in Procedure 3.

**2.3** Name several derivatives of the exocarp, mesocarp, and endocarp.

## Seeds

The seed is a structure formed by maturation of the ovule following fertilization. Seeds are the end products of sexual reproduction, and they house the embryo. Seeds protect, support, and nourish the embryonic plant until **germination** (resumption of growth and metabolic activity). Many plants have developed elaborate strategies to disperse the seeds into the environment (Fig. 12.40).

FIGURE **12.40** Diversity of seeds: (A) burdock; (B) dandelion; (C) maple; (D) touch-me-not; (E) coconut.

### Seed Anatomy

All seeds are covered by a **testa** (seed coat) that protects the seed from drying out, extreme temperatures, bacteria, fungi, and predation. An opening called the **micropyle** is often visible on the seed coat as a small pore. The micropyle allows the pollen tube to enter the ovule to ensure fertilization. In addition, some seeds have a distinct scar, the **hilum**, left on the seed coat when the seed separates from the supportive **funiculus**, or stalk. The nutritive endosperm, beneath the testa, contains copious amounts of starch that nourishes the seed after germination.

Angiosperm and gymnosperm seeds differ in the origin of their stored food. The female gametophyte in gymnosperms provides the food. In angiosperms, the food is supplied by cotyledons. In addition to the endosperm, an **embryo** can be found within the seed. The size of the embryo varies with the plant species. The mature embryo consists of a stem-like axis bearing one (monocot) or two (eudicot) cotyledons. The cotyledons, or seed leaves, are the first leaves to appear in a new sporophyte. They serve as food storage organs for the seedling plant.

Upon examination, a bean has two distinct halves, each a cotyledon, and a corn kernel has a single cotyledon. In monocots the cotyledon may be called the **scutellum**. In monocots the scutellum is highly absorptive. At opposite ends of the plant embryo are the **apical meristem** of the shoot and the root. Many plants have a stem-like axis, the **epicotyl** (Fig. 12.41), with one or more developing leaves above the cotyledon or cotyledons. The resulting embryonic shoot is called the **plumule**. The stem-like portion beneath the cotyledon or cotyledons is called the **hypocotyl** (Fig. 12.41). The embryonic root, or **radicle**, exists at the lower end of the hypocotyl. The radicle and the plumule are enclosed in sheath-like protective structures called the **coleorhiza** and the **coleoptile**, respectively. These structures protect the seed during germination.

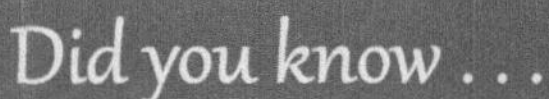

*Did you know . . .*

### How about a Kiss under the Mistletoe?

In Anglo-Saxon, mistletoe literally means "dung on a twig." It was thought to appear through spontaneous generation from bird feces. The etymology does not exactly reflect the romantic reputation of the mistletoe plant! Mistletoe has been considered one of the most magical plants in European folklore. It was thought to bestow fertility and life to people, protect against poisons, and even serve as an aphrodisiac.

The first recorded tradition of kissing under the mistletoe is associated with the Greek festival of Saturnalia and early marriage rites. Mistletoe supposedly brought about fertility and long life. In Norse mythology, two enemies were said to find peace by kissing under the mistletoe. In Victorian England at Christmas, a young lady would stand under a kissing ball of mistletoe and await a kiss. If she was kissed, it could mean romance or close friendship. If she was not kissed, it meant that she would not marry during the next year. Today, the tradition of kissing under the mistletoe appears throughout the holiday season and signifies love or lasting friendship. Ironically, mistletoe berries are poisonous!

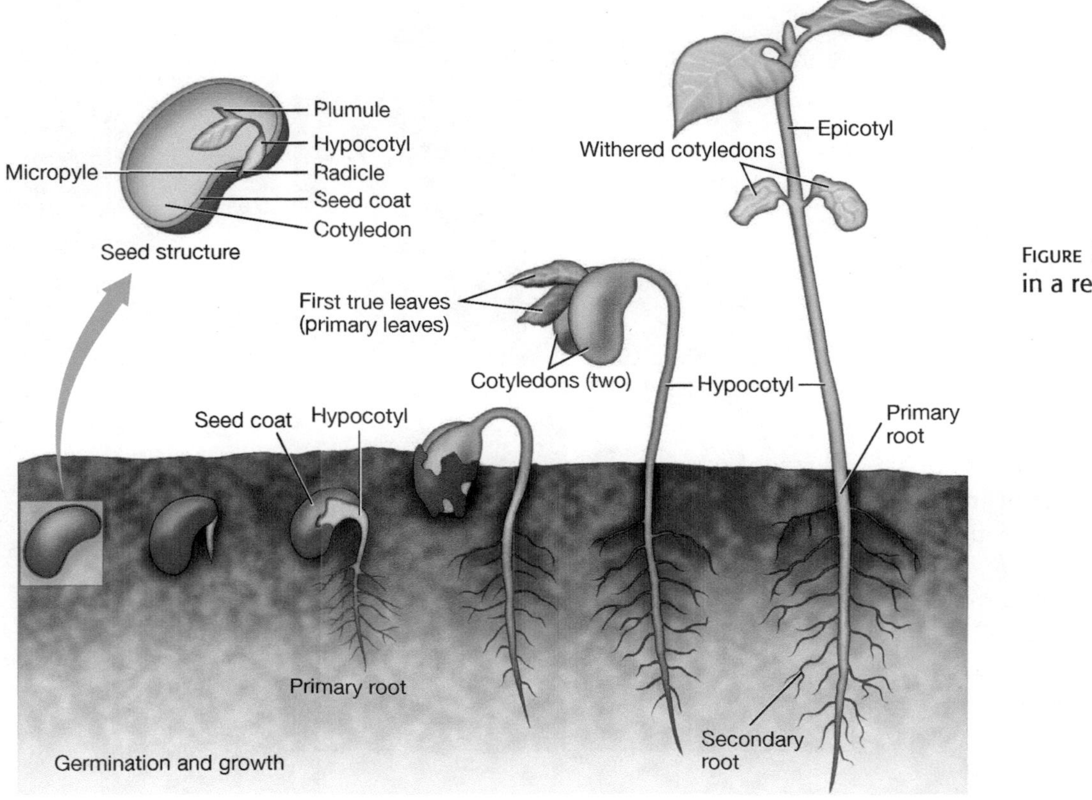


FIGURE **12.41** Germination in a red bean, a typical eudicot.

## Germination

When environmental conditions are favorable, a seed breaks dormancy and germinates, forming a new generation of the plant. The embryos of different species remain viable (capable of germination) for varying periods of time. Seeds of one species of lotus germinated after they were discovered in a 3,000-year-old tomb!

Among the variables influencing germination are temperature, light, water, and scarification, the latter of which can be brought about by bacterial action, stomach acid, freezing, or fire. Usually, seeds germinate while under the surface of the soil. Although soil is the ideal environment for roots, shoots are poorly designed for growth under abrasive soil conditions. Fortunately, nature has provided several mechanisms for protecting the young shoot during its emergence from the soil.

In beans, after development of the root and anchorage in the soil, the hypocotyl grows toward the surface in the form of a hook, gently pulling the cotyledons upward. When the hypocotyl hook reaches the surface, light induces the tissue of the hook to straighten, bringing the cotyledons and the young shoot to the surface. During this process, the plumule is protected between the cotyledons. In peas, the epicotyl elongates and forms a hook that otherwise is similar to the mechanisms of hypocotyl elongation in beans, but the cotyledons remain under the surface. In corn and in some grasses, the coleoptile protects the plumule. The coleoptile is a tough, protective sheath that completely surrounds the plumule. When the coleoptile reaches the surface, light induces it to split, and the plumule emerges.

# Seeds

## Procedure 1
### Iodine and Starch

Why does the clerk mark your $20 bill with that "magic pen"? As you learned in Chapter 2, paper money contains many chemical safeguards to protect it from counterfeiters, including that of removing starch (*amylose*) from the paper. Counterfeiters have not discovered how to remove the amylose, and when counterfeit bills are marked with an iodine pen, the mark appears blue. Gotcha! Iodine reacts with starch, producing a bluish color. In this procedure, the cut surface of a corn kernel is treated with a drop of iodine solution. The storage tissues turn blue because of the presence of amylose.

**Materials**
- ❑ Iodine solution or counterfeit pen
- ❑ Eyedropper
- ❑ Scalpel
- ❑ Plastic petri dish
- ❑ Paper towels
- ❑ Paper currency (e.g., $1 bill)
- ❑ Cotton swab
- ❑ Corn kernel soaked in water for 24 hours

1. Procure the material and equipment from the instructor.
2. Carefully cut the corn kernel in half longitudinally.
3. Place the kernel in the petri dish, and place one drop of iodine solution on the kernel.
4. Wait two minutes, and then describe what happened. What portion of the kernel is blue?

5. Lightly dip a cotton swab into the iodine solution, lightly rub the iodine solution on the paper money, and determine if your "bill" is counterfeit.

## Procedure 2
### Macroanatomy of Seeds

**Materials**
- ❑ Dissecting microscope or hand lens
- ❑ Iodine solution
- ❑ Scalpel
- ❑ Seeds provided by the instructor, such as a red bean, a peanut, a corn kernel, a true rice grain, and a watermelon seed
- ❑ Colored pencils

1. Procure the equipment and seeds.
2. Describe the specimens in detail and, using a dissecting microscope or a hand lens, observe, draw, and label the anatomical features of your specimens in the space provided on the following page (Figs. 12.41 and 12.42). In your description, be sure to include both the plant's common name and the scientific name.

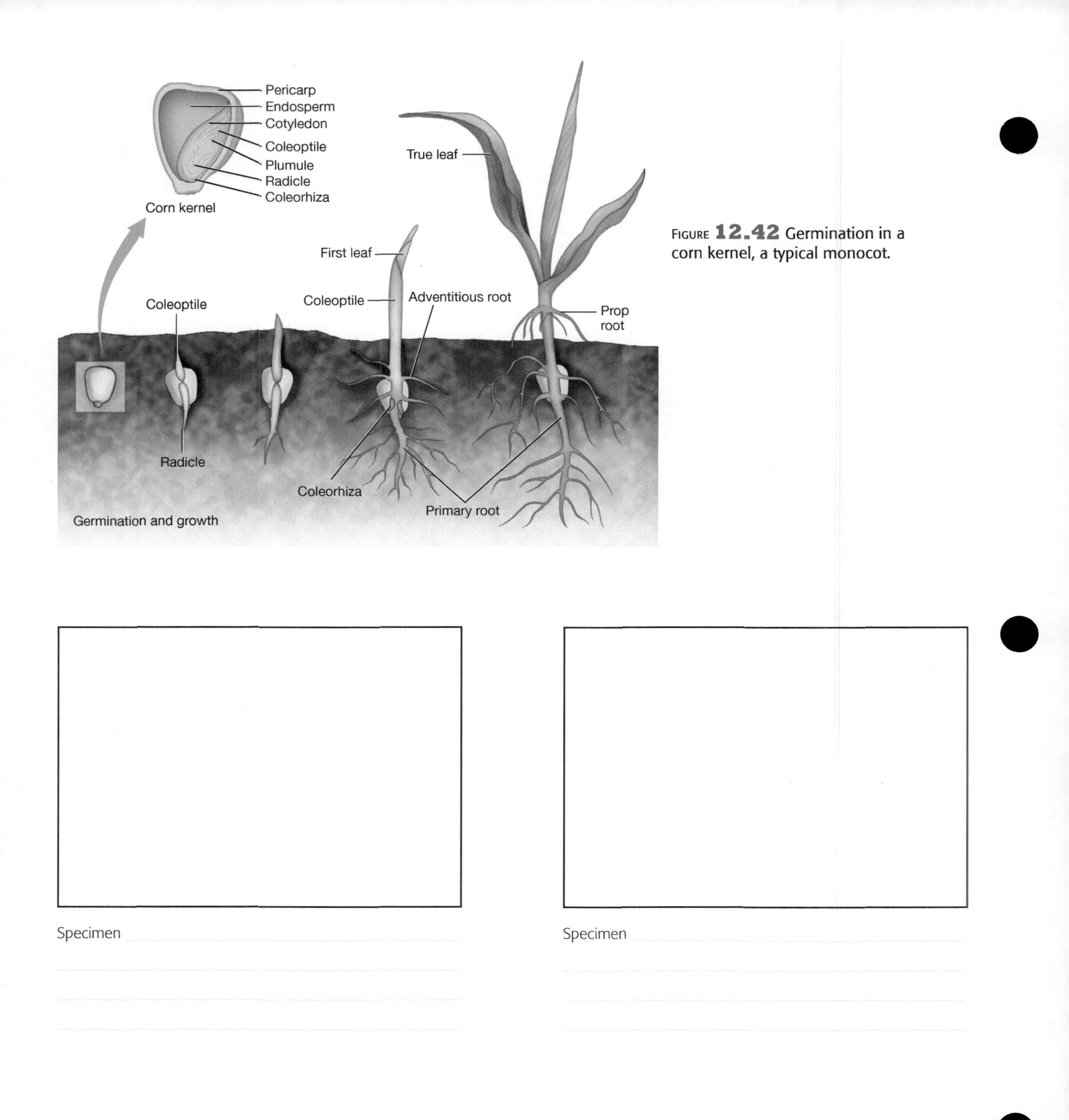


FIGURE **12.42** Germination in a corn kernel, a typical monocot.

Specimen

Specimen

**3** Using a scalpel, carefully cut each seed longitudinally. Describe the specimens in detail and, using a dissecting microscope or a hand lens, observe, draw, and label the anatomical features of your specimens in the space provided.

Specimen

Specimen

## Check Your Understanding

**3.1** What do a counterfeit $20 bill and a corn kernel have in common?

**3.2** What is a cotyledon?

**3.3** What is the function of the radicle?

**3.4** What is the coleoptile?

Name ______________________ Date ____________ Section ____________

## MYTHBUSTING

### Planting a Seed

*Debunk each of the following misconceptions by providing a scientific explanation. Write your answers on a separate piece of paper.*

1. Rosehips (the berrylike fruit structure of a rose) are considered poisonous.
2. Seedless grapes are seedless because they are sprayed with hormones.
3. Flowers were placed on Earth for their beauty.
4. If you swallow a seed, it will germinate in your appendix.
5. The lotus flower provides eternal life.

**1** Label the flower diagram to the right.

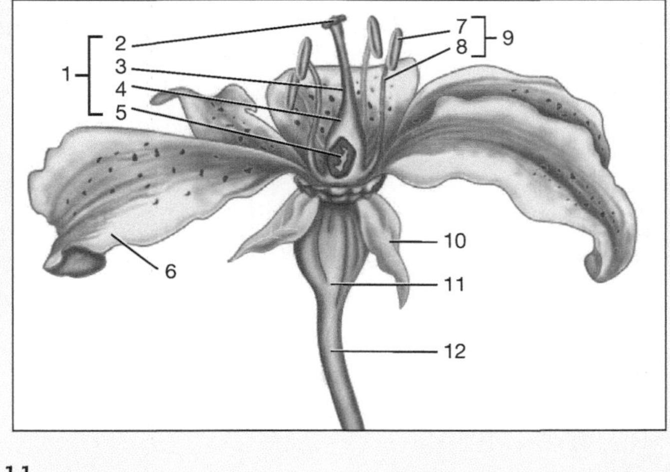


1. ____________
2. ____________
3. ____________
4. ____________
5. ____________
6. ____________
7. ____________
8. ____________
9. ____________
10. ____________
11. ____________
12. ____________

**2** Compare and contrast perfect and imperfect flowers, and provide an example of each.

**3** Compare the germination of a typical eudicot seed and a typical monocot seed.

4 Take an imaginary trip to a market. List 10 fruits you encounter, and indicate the group and type of fruit.

5 Draw a longitudinal section of a tomato, and label the parts.

6 Describe how fruits are classified.

7 What is a peach pit?

8 Describe several means of seed dispersal.

## Review: The Phylogeny of Land Plants

Now that you know about the morphological characteristics of the various groups of land plants, you can use this information to construct a **cladogram** that describes the **phylogeny** of the plant groups that we have studied.

**What is phylogeny?** This is the evolutionary history of a species or group of related species.

**What is a cladogram?** A cladogram is a type of phylogenetic tree, a diagram that represents the evolutionary relationships for groups of organisms. The cladogram depicts a hypothesis on the sequence of changes (= branching events) that occurred to generate a given set or group of present-day organisms from their common ancestor. Each branch in the tree of life represents a **clade.** Clades include what is thought to be the source taxon and all its descendants.

While this cladogram will be constructed based on morphological characteristics, there are other characteristics that help scientists to understand the evolutionary relationships of plants. DNA sequencing, particularly of the circular genome of the plant chloroplast, has provided useful information into the phylogeny of land plants.

| | Cuticle absent (0) present (1) | Xylem & phloem absent (0) present (1) | Pollen absent (0) present (1) | Seeds absent (0) present (1) | Flowers absent (0) present (1) | Fruits absent (0) present (1) |
|---|---|---|---|---|---|---|
| Green algae | | | | | | |
| Mosses | | | | | | |
| Ferns | | | | | | |
| Gymnosperms | | | | | | |
| Angiosperms | | | | | | |

Now to construct the cladogram. The green algae is an **out-group,** a group that is known to have diverged before the group we are studying, the Embryophytes or land plants, the **in-group.**

**Green algae will be the starting taxon of our cladogram. Label letter A as "green algae."** They contain the same photosynthetic pigments as land plants, chlorophyll *a* and *b* and carotenoids, use starch as a food storage molecule and have cell walls are made of cellulose, also like land plants.

What characteristic do members of all the in-groups share (but the out-group lacks)?

________________________________________________________.

This marks your first branch on the cladogram. **Write this trait in box b.**

Now consider the next characteristic – the presence of xylem and phloem. This marks your next branch on the cladogram. **Write this trait in box c.**

Which of the in-group **lacks** this trait?

________________________________________________________

**Label this group as letter B.**

Continue with this pattern to complete your cladogram. Note that more than one trait may appear in a box that marks a branch on the cladogram.

## Cladogram of the Phylogeny of Land Plants

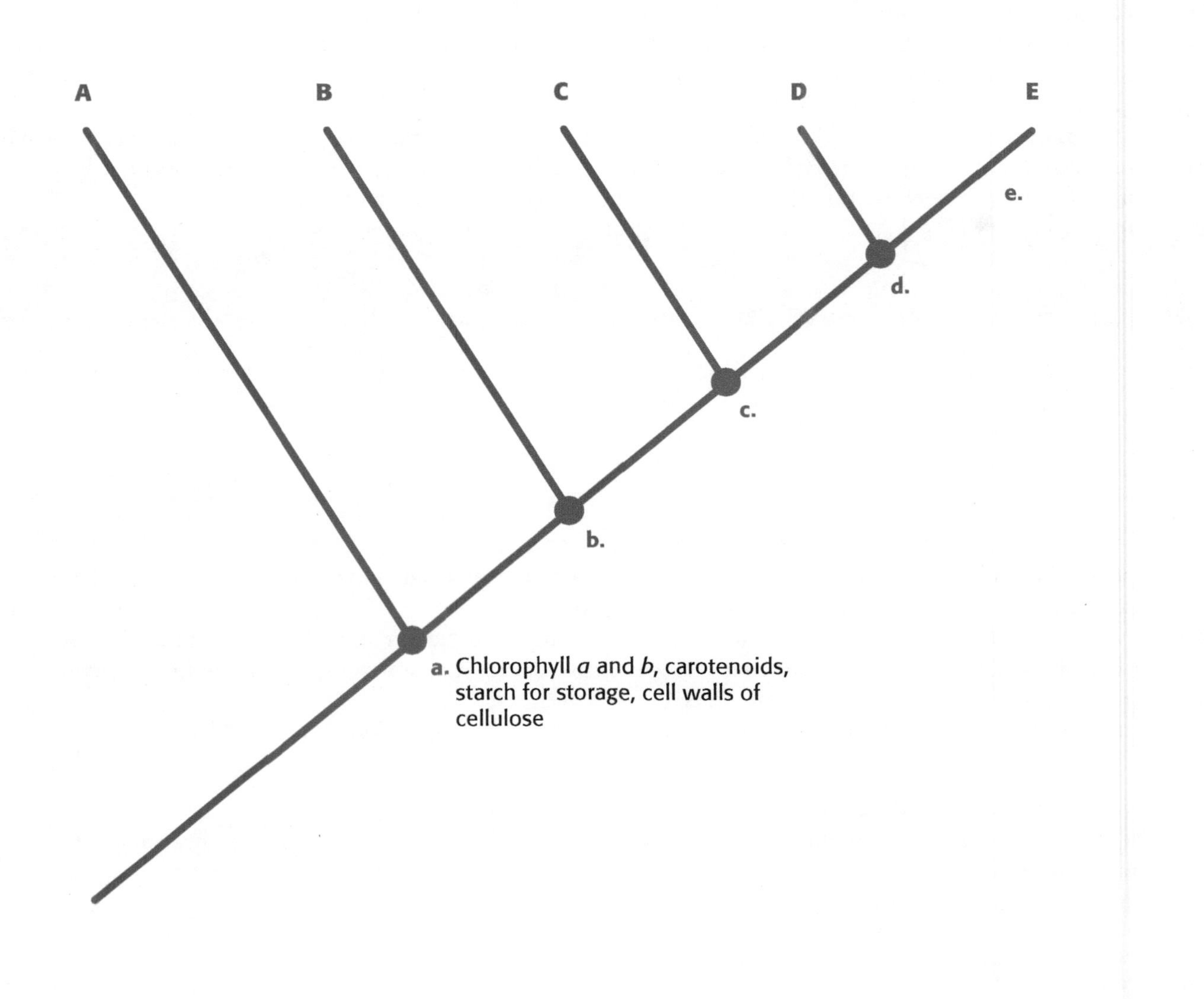

# Lab Practical Final

Protists and Fungi
Photosynthesis
Bryophytes and Ferns
Angiosperms and Gymnosperms